Eswaran Subrahmanian · Toluwalogo Odumosu
Jeffrey Y. Tsao
Editors

Engineering a Better Future

Interplay between Engineering, Social Sciences, and Innovation

Editors
Eswaran Subrahmanian
Carnegie Mellon University
Pittsburgh, PA
USA

Jeffrey Y. Tsao
Sandia National Laboratories
Albuquerque, NM
USA

Toluwalogo Odumosu
Engineering and Society
University of Virginia
Charlottesville, VA
USA

https://doi.org/10.1007/978-3-319-91134-2

Preface

Engineers make up roughly 2% of any country's population, however, they exert an outsize impact on the lives of citizens and on society in general. Since the Industrial Revolution through the current Information Revolution, engineers through their work have played the roles of sociologist, economist, political scientist, and policy analyst, changing the patterns of our lives. Socially embedded innovations such as the cell phone and the Internet are transforming the economic, social, cultural, and physical well-being of people all over the world. The debate is often over how the technology fits into society and not necessarily over how society is reflected in the technology developed, i.e., for example, how the technological system reflects preexisting bias and inequality. As the 1991 NAP report on "Engineering as a social enterprise" states, the dominant view of engineering is one of detached technological quests apart from society. The social implications of engineering and the social influence of engineering is often left out leading to a paucity of interaction between social sciences and engineering in any meaningful way. Engineers are often unaware of technological history and the role of social forces in the history of the development of technologies. Furthermore, the importance of social understanding of innovation processes is often downplayed both in daily and academic discourse on engineering. The workshop was designed to bring together these two cultures. Our hope is that they are not as divided as C. P. Snow declared, but rather, it is the lack of recognition of the intertwined nature of engineering and social sciences that prevents interactivity.

The goal of the workshop was to bring together engineers and social scientists to examine the role that engineers have in the future of our living environments and our economies, and to begin a discussion on how best to integrate the social sciences into engineering practice and research. The best innovation systems anticipate characteristics of technological use and culture, and these are best informed by the social sciences. The symposium addressed the role of social science research in shaping engineers' view of their work and its role in the transformation of society.

Engineering is shaped and informed by social sciences in ways that would benefit from open discussion and greater integration. Often the larger population and engineers themselves are seldom aware of engineers' contributions to social sciences. From Stevenson to Ford, engineers continually envision and actualize profound changes to social character and behavior. Individuals such as Benjamin Whorf, Vilfredo Pareto, and Fredrick Taylor whom, while engineers in practice, made significant contributions to social sciences, suggesting the close interrelationship between engineering and social sciences. Given the range of problems faced by global society from global warming, the automation of work, to environment degradation and social inequity, it is clear that only through the collaboration and appreciation of the each other can engineers and social scientists work together to solve these critical problems.

The workshop brought together a variety of scholars from the social sciences and engineering (see list below) to Carnegie Mellon University for a period of 2 days. The workshop was divided into three themes: Meeting at the Middle (challenges to educating at the boundaries); Engineers Shaping Human Affairs; and Engineering the Engineers (thinking about design in designing thinking).

Pittsburgh, USA — Eswaran Subrahmanian
Albuquerque, USA — Dr. Jeffrey Y. Tsao
Charlottesville, USA — Prof. Toluwalogo Odumosu
Organizing Committee

Acknowlegments

We would like to thank Prof. Venkatesh Narayamurthi for his encouragement to find ways to make this workshop happen. He brought us, Subrahmanian, Jeffrey Y. Tsao, and Tolu Odomosu together, as organizers that allowed us to pull off the workshop and the edited book. We would also like to thank, Prof. Maryann Feldman, Heninger Distinguished Professor in the Department of Public Policy at the University of North Carolina, who was the Program Director of SciSIP program of NSF, supported and funded this workshop. Our thanks to Prof. Douglas Sicker, Head of the Department of Engineering and Public Policy for his active participation in the workshop, Prof. Burcu Akinci, Associate Dean for Research at College of Engineering, Rebecca Gray, and all other Staff at the Engineering Research Accelerator at CMU who supported in making this workshop successful. This workshop was funded by the NSF grant 36513.

Contents

Contributors

Georges Amar Mines ParisTech-PSL Research University, Paris, France; RATP, Paris, France

Glory E. Aviña Sandia National Laboratories, Albuquerque, NM, USA

Travis L. Bauer Sandia National Laboratories, Albuquerque, NM, USA

W. Bernard Carlson Engineering and Society Department, University of Virginia, Charlottesville, VA, USA

Joel Chan School of Information Studies, University of Maryland, College Park, Maryland, USA

George W. Crabtree Argonne National Laboratory, Argonne, IL, USA

Steven P. Dow Human–Computer-Interaction Institute, Carnegie Mellon University, Pittsburgh, PA, USA

Ron Eglash Department of Science and Technology Studies, Rensselaer Polytechnic Institute (RPI), Troy, NY, USA

Gregory J. Feist San Jose State University, San Jose, CA, USA

Armand Hatchuel Chair of Design Theory and Methods for Innovation, MinesParisTech-PSL Research University, Paris, France

Paulien Herder Department of Engineering Systems and Services, Faculty of Technology, Policy and Management, Delft University of Technology, Delft, The Netherlands

David Howarth University of Cambridge, Cambridge, UK

Curtis M. Johnson Sandia National Laboratories, Albuquerque, NM, USA

Ade Mabogunje Center for Design Research, Stanford University, Stanford, CA, USA

Venkatesh Narayanamurti Harvard University, Cambridge, MA, USA

Toluwalogo Odumosu University of Virginia, Charlottesville, VA, USA

S. Thomas Picraux Los Alamos National Laboratory, Los Alamos, NM, USA

Yoram Reich School of Mechanical Engineering, Tel Aviv University, Tel Aviv, Israel

R. Keith Sawyer University of North Carolina, Chapel Hill, NC, USA

Christian D. Schunn Learning Research and Development Center, University of Pittsburgh, Pittsburgh, PA, USA

Richard P. Schneider glo-USA, Sunnyvale, CA, USA

Dan Siewiorek Human Computer Interaction Institute, Carnegie Mellon University, Pittsburgh, PA, USA

Austin R. Silva Sandia National Laboratories, Albuquerque, NM, USA

Neeraj Sonalkar Center for Design Research, Stanford University, Stanford, CA, USA

Jennie C. Stephens School of Public Policy & Urban Affairs, Northeastern University, Boston, MA, USA

Rickson Sun IDEO, Palo Alto, CA, USA

Shyam Sunder Yale School of Management, Yale University, New Haven, CT, USA

Jeffrey Y. Tsao Sandia National Laboratories, Albuquerque, NM, USA

Ana Viseu Universidade Europeia, Lisbon, Portugal; Centro Interuniversitário de História das Ciências e Tecnologia, Universidade de Lisboa, Lisbon, Portugal

Margot Weijnen Department of Engineering Systems and Services, Faculty of Technology, Policy and Management, Delft University of Technology, Delft, The Netherlands

Jameson M. Wetmore School for the Future of Innovation in Society, Arizona State University, Tempe, AZ, USA

Langdon Winner Department of Science and Technology Studies, Rensselaer Polytechnic Institute, Troy, NY, USA

Introduction

Background

Engineering a Better Future: Interplay between Engineering, Social Sciences and Innovation, a workshop funded by the National Science Foundation, was held from April 15–16, 2016 at the Carnegie Mellon University. The workshop was held at a time in history when innovation and design thinking "fever" are at a high pitch, when engineers are "king", and when the products of engineers—technology, is shaping our lives, social relationships, habits, and even needs, in a kind of uncontrolled, technology-centered evolution of our society.

We are not always happy with the shape that this uncontrolled evolution takes. Overwhelming problems of day-to-day life have not yet been alleviated by technological systems, as evidenced by the sub-par quality of life of billions on the planet. In other cases, technological systems have been the *cause* of problems. As discussed by Galbraith (1998), new needs and desires in an affluent, technology-rich society can be created where none previously existed. There are also other extreme cases such as the whitening cream sold to women in India who have become preoccupied with obtaining lighter complexions, or the burning need to own the next iPhone in order to keep up with peers. These kinds of consumption patterns are global and are deeply ingrained in our psychological makeup, i.e., our fears and personal anxiety at being left out of the game of life.

The question we should ask is: what would "engineering a better future for humanity" look like? One possibility, the metaphor of society as a machine, conjures authoritarian order and a socially engineered, "designed" society, which we by instinct reject. Such social engineering would appear to obstruct and limit the infinite possibilities that humanity might aspire to. The other extreme possibility, the metaphor of society as a complex adaptive system, conjures a runaway and perhaps ultimately self-destructive mix in which humanity, whose instincts evolved and adapted in prehistoric times, interacts with evermore powerful technologies provided to it by willing but oblivious engineers.

Are there middle possibilities, in which we avoid these two extremes through meaningful ways of socializing engineering and engineering the social in a collective sense? Can a better *joint* understanding by engineers and social scientists of societal values, understanding sociopolitical contexts, ethical education, and empathy for the environment, enable to us to avoid the blindness that might lead to self-destruction? For example, in the 1991 report "Engineering as a Social Enterprise", the last National Academy of Engineering Publication on this subject (Sladovic 1991), Thomas Hughes compared two different solution approaches to electrification, one in the US and the other in Europe, and showed the intricate interrelationships between society and engineering. Or, for example, Subrahmanian taught a course on engineering problem formulation in different social contexts at CMU. The course was extended with Prof. Paulien Herder from the Technical University of Delft, Netherlands to illustrate the social and cultural context of the two countries in their approach to dealing with engineering problems, and how the choices made in each of these societies led to very different solutions (for example, in transportation modes) (Subrahmanian et al. 2003). In other words, can a better interplay and mutual understanding between engineers and social scientists help us shape technologies and the social contexts of their uptake in ways which better serve humanity? We are fully aware of the challenges associated with finding the right balance within these middle possibilities.

First, technologies, the products of engineers, are complex and their societal ramifications are difficult and sometimes impossible to predict. Recently, there have been public discussions about smart cities, and the imaginative reengineering of the environment and the connected lives that are possible in the era of the networked world. This is no different from the imaginative reengineering of transportation systems (cars), the unintended consequences of which gave us suburbia and long commutes. It is also no different from the imaginative reengineering of information technologies which are intrusive and paradoxically, make the society both more transparent and less transparent at the same time. No matter how socially conscious or aware of social science principles an engineer might be, his/her ability to forecast how technologies will interact with society is limited. Technologies can both liberate and imprison us, both in surprising ways. Whether we imagine an extractive economy (state or private), a generative circular economy (Eglash in this volume), or a neoliberal model that promotes consumption through its mythology of innovation (Winner, in this volume), how the products and services that are created integrate and transform human society is difficult to anticipate.

Second, it is not just technologies that can be engineered. As Howarth (in this volume) argues, legal systems are also engineered and have a life cycle. We engineer laws to deal with technologically engineered products such as emissions and corporate finance laws that also change society in specific ways. Furthermore, if law is "engineered", isn't economics engineered too? If it is true that "economics is an artificial science," then surely, economics should be understood as being subject to principles of design and engineering in similar fashion. Nobel Economists Ostrom and Roth use design and engineering in the way they approach problems in economics (Ostrom 2005; Roth 2002). Varian (2002), another theoretical economist, made the case that economists are more and more, being asked to be engineers. This recognition of the "engineering" of our social lives beyond traditional understanding of engineered physical technologies brings to the fore the importance of a necessary interplay between engineering and the social sciences.

Third, engineers of physical technologies and engineers of social technologies and human affairs are human, all of them, and are just as prone to failures of integrity and behavior as all humans are. Veblen (1963), the American sociologist and economist, argued that the Engineer was the engineer of price and believed that the engineer would be the revolutionary in changing society. That did not happen. In fact, engineers in America have been subordinated by their managers. Charles Perrow, who could not attend the workshop due to illness, in his request to the workshop participants asked (Perrow 2016):

> The most important question I would broach at the conference is why are mid and lower level engineers willing to run great risks that are ordered from the top? In my study of major accidents and disasters and fraudulent organizational behavior, I often find that engineer's warnings had no effect upon the top, and I assume they went ahead with great reluctance, fearing sanctions. If that is the case the problem is with top management (who are often engineers of course) and we should be studying them, and not expect lower level engineers to be heroes.

This request raises the question: when an engineer becomes a manager, does he/she give up his/her engineering ethics of safety first with the products that are built and operate so as instead to satisfy shareholders? Even though Veblen (Veblen 1963) put his faith in the engineer, he was aware of the rise of corporate finance and its control of the engineer (Veblen 1963, p. 55), similar to how Edward Layton details the weak power of the engineers in his book on the "Revolt of the Engineers" (Layton 1986).

Of course, we do not know the answers to these various challenges, and the purpose of the workshop was to explore how, through an interplay between engineering and the social sciences, answers might emerge. Our hypothesis is that both engineering and social science can enrich each other with their methodological approaches and insights, and ultimately, can help "engineer a better future."

On the one hand, we encourage social scientists to expand their scope to include how engineers think and work, and to develop new principles of engineering that apply not only to physical things but also to human societies. Just as science, cinema and arts influence our thinking of the future, social sciences need not just study universal social science principles, they can also study social structures and rules that work in practical contexts (Sundar, this volume). For example, Ostrom (2005) in her work on institutional analysis and development frameworks creates a grammar and a framework to describe organizations that manage common pool resources. The creation of this grammar is similar to what Redtenbacher did for machine design in Germany in 1850s to promote economic development as it was lagging behind France and Britain. Redtenbacher's engineers were not just technicians, they were expected to be versed in ways of the society, humanities, and the arts as well (Wauer et al. 2010). This is the same message that one sees in the work of Vitruvius (Pollio 1914), the author of the Ten Books on Architecture.

On the other hand, we encourage engineers to better understand how their work influences human society and how the way they work is itself governed by social principles. This latter idea is in our opinion, important but under-appreciated. Social philosophy, attitudes, capabilities, and institutional mechanisms all play a role in the way in which engineered products are developed. The work of Hausman et al. (2014) on the economic complexity of nations shows how the level of complexity of products produced by a country is directly related to its skill base and institutions. Societal mechanisms and attitudes have a clear relationship to the kind of engineering that can be, and is done (product/process complexity and sensitivity).

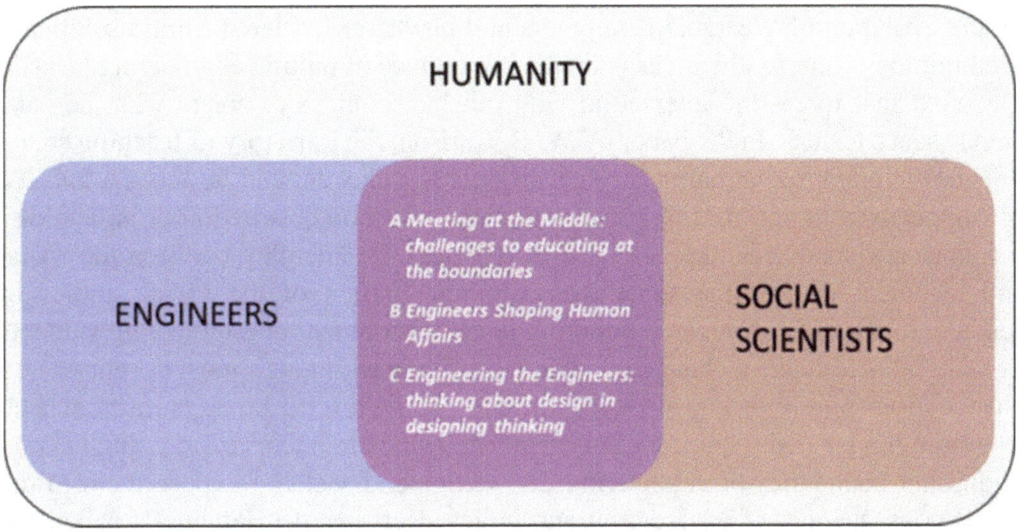

The workshop itself was not sufficiently comprehensive. It would not be possible in a single workshop to illustrate all the possible ways social science disciplines interconnect with engineering. However, it brought together a variety of people from different disciplinary backgrounds, all of which are well known to have ventured into this middle space rarely visited by most engineers or social scientists. We hope in bringing these individuals together, and in publishing their thoughts in this volume, we help catalyze additional serious conversation between the two communities.

To focus the workshop, it was organized along the following three themes as shown in the figure: Meeting at the Middle (challenges to educating at the boundaries); Engineers Shaping Human Affairs; and Engineering the Engineers (thinking about design in designing thinking). In the remainder of this Introduction, we offer a brief summary of these three themes, along with the various papers associated with them. The breadth of this collection is not indicative of the entire range of possibilities of interplay between social sciences, engineering, and innovation, but rather instead, a representative sampling.

Meeting at the Middle: Challenges to Educating at the Boundaries

The collection of papers in this theme is mainly about incorporating social sciences into engineering curricula in various universities in Europe and the US. In the case of The Technology Management program at Technical University of Delft, the choice of "Next Generation Infrastructure" as the focal topic allowed for different disciplines to come together and study problems from perspectives of the other disciplines. In Jennifer Stevens' work on engineering and environment, the focus on climate change provides the common object of interest. In Dan Siewiorek's paper on the evolution of Wearable Computers and his "User Centered Multidisciplinary Methodology" course, the focus is on the importance of putting engineering tasks in a context that forces the interaction with other disciplines to create what he calls "renaissance teams". In the paper by W. B. Carlson, the trajectory of teaching social sciences to engineers at University of Virginia is followed, and he seeks a balance of perspectives in not making engineering into a social construction or an apology for what engineering is: the goal is to understand the interplay between the social and technical without the sometimes-pejorative jargon of the social studies of science. In the case of Jamie Wetmore, the disillusionment of engineering students in being able to make a change in society is addressed: though most do not want a change from purely technical education and are willing to be cogs-in-the-wheel. Wetmore has created a space for those students, who are interested in going beyond traditional boundaries of engineering knowledge and seeks to expose them to the social consequences of engineering and policy decisions. In Hatchuel's paper, the history of the evolution of engineering studies at Mines ParisTech, France, is reviewed over three major periods: from its beginning as a technical school; to a science-based generalist school with the inclusion of management, political science, and law as part of the curriculum; to the third period where creative, critical, and socially responsible thinking are emphasized in the context of globalization and financialization of engineering and economics.

Technology, Policy and Management: Co-evolving or Converging?

In "Technology, Policy and Management: Co-evolving or Converging?" Margot Weijnen and Paulien Herder trace the logic of the faculty of Technology, Policy and Management (TPM) from its origins in 1992. The primary focus of the faculty is what they term "comprehensive engineering", where the goal is to combine insights from engineering sciences, social sciences, and humanities. The premise is that this combination allows for the exploration of interfaces between systems, governance, and values, especially for a networked urbanized society. The program's evolution from systems engineering to "comprehensive engineering" was motivated by the Next-Generation Infrastructure (NGI) project that created an environment to explore interdisciplinarity in all its richness. The project also articulated the idea of comprehensive engineering as the modeling and design of socio-technical systems that need to mediate conflicting public values. The chapter then traces the history since 1992 of this move and its effects.

The main point of this chapter is that engineering of infrastructure is not new, but infrastructure and its governance structure often evolve independently. Moreover, even though most infrastructures evolve independently of each other, within an infrastructure domain such as the railway infrastructure, telecommunications, or power, which cross national boundaries, the interrelationships between infrastructures must be recognized and the need for new science of socio-technical systems becomes imperative. It is in this context that the authors argue that TPM had to move from systems engineering to socio-technical systems engineering where methods and models integrating for values, governance, and systems became paramount. New methods such as serious gaming and agent-based modeling that allowed for hybrid models of engineering and social sciences were developed. The move was important for this program to integrate the training in social sciences, engineering, and modeling of complex hybrid systems and has been critical to educate a new class of engineer. The educational program at TPM has embraced this view fully and has been producing Ph.D.s and masters students, who are able to treat social sciences and public values not just as context variables but integral to the design of such systems. This approach has allowed for the program to recognize that systems, governance, and values need to converge.

Evolving from Single Disciplines to Renaissance Teams

The chapter by Dan Siewiorek from Carnegie Mellon University, "Evolving from Single Disciplines to Renaissance Teams", traces the motivation and history of a course that started as the creation of a single-board computer to the development of wearable computers. This chapter traces the evolution of the course by identifying the failures of the single-board computers that were designed purely by electrical and computer engineers to the outreach to other disciplines. They found their inspiration in the work of the late Randy Pausch, who had created a multidisciplinary course at CMU as part of the entertainment technology master's program with the explicit goal to create what he termed as "Renaissance" Teams. The notion here is that no one person can possibly cross enough disciplines to be a "Leonardo" of current science and engineering, but one could create teams that mimic the same attitude and openness needed to solve complex problems by transcending technology to address the needs of the user. The chapter traces the characteristics of this course that has as its cornerstone a User-Centered Multidisciplinary design methodology. They also used of documentation tools for understanding the interaction between the disciplines and groups through natural language processing of documents. The conclusion describes the lessons learned from this continuing exercise.

Knowledge, Skill, and Wisdom: Reflections on Integrating the Social Sciences and Engineering

The chapter, "Knowledge, Skill, and Wisdom: Reflections on Integrating the Social Sciences and Engineering", by W. Bernard Carlson, from University of Virginia (UVA) traces the history of the integrative social sciences in the undergraduate engineering curriculum at UVA. Carlson begins the paper by making the case that there exists an inherent tension between dependency on technology and ignorance about it in the general populace versus the continued specialization that haunts the engineer from understanding the connection to society of their work. In addressing this problem, he identifies three important components of engineering education: Knowledge, Skill, and Wisdom which he makes the grounding for creating reflective practitioners of engineering in the sense that was first described by Donald Schon. In UVA's composition of the three components: Knowledge is the use of equations, methods, and other modeling techniques; Skill is the know how to use models, theories, methods, and technique as the they are all about use and manipulation of information; and Wisdom is the skill to know when and where to use these skills while acknowledging that not all is captured in a mathematical model or a theory.

Students are also exposed to the need to understand their limitations and go beyond them to recognize the scope and limitation of their creations from moral, ethical, and socially responsible perspectives. In this chapter, Carlson lays out the history of the evolution of this philosophy of engineering education arising out of a peculiar history of engineering at UVA from a course of writing, presentation, and communication to a four-course structure. The students learn writing, presentation communication as one also learns to think about the interactions between society and technology and its in a nuanced way varied implications in a nuanced manner. To do so requires the evaluation of positive and negative consequences and the ability to think about unintended consequences and their potential impacts. This is done by incorporating case studies, reading in science technology studies and bringing all of them together in a final thesis project. Carson in his chapter also alludes to the challenge of teaching the social aspects of design without the jargon of social studies in technology in the context of production of knowledge. Finally, Carson points to an attitude that is critical to teaching this perspective to the students, i.e., it neither should be an apology for what engineering is nor should it make all of engineering exclusively a social construction.

Innovations in Energy-Climate Education: Integrating Engineering and Social Sciences to Strengthen Resilience

The fourth chapter in this collection by Jennie C. Stephens explores the opportunity afforded by the energy-climate nexus to incorporate social sciences in engineering education. Her primary thesis is that the energy-climate nexus not only transcends silos of disciplines but also connects to practice through its impact on behavioral, economic, and social well-being in communities small and large. Thus, it creates a space for students with diverse interests and backgrounds to relate engineering to their lives. Beyond engineering, it provides a substrate for STEM education in K-12 at the precollege level. She also makes the case the that the rise of renewable energy and the distribution of energy resources could lead to more democratic forms of governance in resources and energy at the local, regional, and national levels than a distribution dominated by traditional energy resources. She explores the concept of an Energy Democracy that strives to integrate energy policy and social policy to rearticulate energy systems as distributed systems that accrue social benefits to local communities out of the reach of powerful energy interests. She makes the case that the climate-energy nexus brings together the political, social, and technical in a fashion that allows for innovation. Innovation occurs not just in the local communities but in the nature of engineering education itself. By its embrace of the social sciences in engineering through direct engagement, project-based and contextualized learning of engineering would create new career paths and social engagement.

Reconnecting Engineering with the Social and Political Sphere

The chapter by Jameson M. Wetmore, on "Reconnecting Engineering with the Social and Political Sphere", narrates his experience and solution to the problem of loss of interest in engineering in students who come to the field thinking they are going to be changing the world or doing something good for the world. His observation is that most engineering students feel comfortable being "cogs-in-the-wheel" when asked about going beyond the technical confines of their work. Wetmore's question is on the means of transforming engineering students' perception and role as "cogs-in-the-wheel" to be capable of addressing broader challenges in society using Callon's image of an engineer as an "engineer-sociologist". In a way, he is appealing to those few who are yearning for that broadened experience by engaging them in challenging problems in the world. Wetmore describes two efforts that he is conducting at ASU. One is called "Science outside the Lab", where science and engineering Ph.D. students are brought together with policymakers who work at the interfaces of science, policy, regulation, lobbying, and others in the government over a 2-week workshop. The second lab is based on community experience and involves engaging with the community in a developing country. In these projects, the goal is to dissuade students from starting with a technological solution to rethinking the methodology to include decentering the technology, listening to and learning from the community and empowering the community. The program has been largely successful in getting engineers to think in ways they did not before is evidenced by the careers they pursue. While they were not converted to engineering sociologists they became cognizant of social, political dimensions in engineering and that they can explore this gray space thorough bridges to other disciplines.

Ecole des Mines de Paris: A Few Lessons from a Long History

The final chapter in this collection, "Ecole des Mines de Paris: A Few Lessons from a Long History", by Armand Hatchuel, traces the history of the school through three periods of evolution as a professional training school (1815–1890), subsequently as a generalist school (1890–1967) and to the inclusion of a research-based school (1967–2014) with the additional distinct lines of research pursued in dedicated centers. This history is based on the evolution of the pedagogical approach and identity of an engineer over time.

The goal of the professional training school (1815–1890) was to teach "The art of mining" that inculcated a critical function to restructure the mine operations and practice to enhance poor management of mines and treatment of miners. Beyond courses, in the basics related to metallurgy, materials, and crystallography, the goal was to connect to practice through a course on mine and machine operations. This course served as the basis for field studies that led to documenting mining practices all over the world. After 1850, developments in science including thermodynamics and the science of machines required a transformation of the pedagogy to incorporate training in mathematics, physics, and chemistry training as prerequisite. With these additions, the identity of the engineer had changed from a technician to an "Applied Scientist" resulting in the higher social status of an engineer. There was also a recognition of the engineer's role in the "creative" act to enhance the well-being of people and the wealth of society. Changing social environment and institutions led to an addition of a course on "Legislation and Industrial Economy", a generic space of context in the form of legal responsibilities from an engineering perspective.

The second period of the school (1890–1967) moved toward a "Generalist school" and the pedagogical structure was changed to accommodate the new scientist and engineer/manager dual identity. During this period with the rise of industrial chemistry and electricity, the school embraced industrial chemistry and integrated electricity into machine operations courses. It made changes to law and industrial economy course and added many other courses to incorporate industrial science of management. The school in the process of creating the new identity created a new pedagogical structure with general scientific education including mathematics, chemistry, and physics, training in generic technologies including management, economics, and sociology and in a chosen specialization that were available within the school. In the last period 1967–2014, there was the rise of a "Research Based School" that absorbed the "Generalist School". This period saw the incorporation of a doctoral program and new research centers that covered economics, sociology, and management among other fields. Now, the new identity was one of a designer, innovator-scientist, and scientist-entrepreneur. The paper concludes by projecting the future of the education of an engineer with an identity that is critical, creative, and socially responsible. Increasing financialization of the economy, social and economic inequality, and environmental degradation calls for a new breed of engineers. The future students should be contributing to collective progress though new forms of responsible firms and industry that reject and regulate the abuses of globalized firms through legislative frameworks and standards.

Engineers Shaping Human Affairs

The collection of papers in this second section highlights the interplay between social sciences and engineering and vice versa. The papers are from different perspectives by people from very different backgrounds. Ideas explored include (a) an exposition by as social studies of science scholar of the cult of innovation in today's science and technology discourse and its corrupting influence in promoting a neoliberal agenda that primarily serves the elite, (b) an alternative model of engineering and development from a social scientist/systems scholar to a move away from the model of exploitative value generation to circulation of value within a community through generative engineering, (c) a call from an engineer/polymath for a conceptual prospective thinking rather than predictive thinking as the future is not simply a continuation of the past, (d) a legal scholar's use of models of rate of failures in engineered objects over the life cycle to show that laws wearing out over time have similar patterns to engineered products wearing out over time, (e) an engineer's call to ask the question of who, why, what, and how as fundamental questions to ask about the role and goal of artifacts and their political, technical, and social goals in designing our future; and (f) a management scholar questioning the role social sciences can play in engineering given their lack of engineering knowledge and the difficulty of normative social science methods to guide technological developments and understand possible paths for their interaction.

The Cult of Innovation: Its Myths and Rituals

Langdon Winner his chapter on, "The Cult of Innovation: Its Myths and Rituals", designates "innovation" as a "God term" in the same vein as the terms revolution, progress, and frontier. The power of god terms comes from the conceptually perceived sense of infinite possibilities of betterment or embedded "inherent potency". For example, how "innovation", the term, has become so pervasive that entire corporate, academic, and other institutions have some form of programs, centers, and courses focused on innovation. The primary argument of this chapter is that most of what is celebrated as invention are gadgets or things like apps associated with gadgets that serve mainly the elite, even though there are attempts at making innovation social or for directing it to sustainability and broader social goals. Winner, while acknowledging the idea of innovation as something that has a past and shapes the future, decries the concept of disruptive innovation. This idea of innovation as disruption and destruction to create new order moves away from planning or recreates new institutions that are about profit and individual "innovativeness" as a means of social order. This model of innovation that is operative is based on the hope of that technological future that will solve global problems while questions of the common good are cast away. Winner coins the term "Procastovation" to signify the misguided nature of future technological fixes through innovation as the solution to global problems of poverty, climate change, and inequality while delaying implementations of solutions that are available today. He argues that there are alternative ways to create broader discourses as was done in a number of cases in the past to address the larger common good. However, Winner argues that the current cult of innovation is a jewel in the crown of neoliberalism with its emphasis on innovation as the discriminating factor in the future of occupational survival of individuals, societies, and companies that could only lead to at best messy and at worst catastrophic outcomes.

A Generative Perspective on Engineering: Why the Destructive Force of Artifacts Is Immune to Politics

Ron Eglash, in his chapter, "A Generative Perspective on Engineering: Why the Destructive Force of Artifacts Is Immune to Politics", proposes a counter model to the model of engineering design that is not based on exploitation of value to one that is based on maintaining and circulating the value without alienation through the commons. He proposes a model of ethics and justice, "Generative Justice", in the production of artifacts that is sensitive to the three values of labor, ecological, and expressive. He distinguishes the latter two in terms of the value associated with ecological preservation, while expressive value is associated with the free ability to express thought culturally and individually. There is a clear emphasis of circulation that is critical for sustaining these values with in a community. He defines generative justice as the universal right to generate unalienated value and directly participate in its benefits; the rights of value generators to create their own conditions of production; and the rights of communities of value generation to nurture self-sustaining paths for its circulation.

There are two further hypotheses he explores through his case studies. The first is the contention is that both market capitalism and socialism (state capitalism) are based on value exploitation with the same consequences of pollution, environmental degradations, and alienation of labor. The second is that an extractive model as the driver of production of artifacts irrespective of politics has the same consequences in terms of alienation. He supports this hypothesis through case studies of two short-lived experiments of worker inclusion from an expressive value perspective in both market capitalist and state capitalist countries that were not sustained and went back to the model of top-down extractive model of production. He then elaborates a path through a number of case studies including use of crowd-sourced Arduino's, watchmaking and teaching of mathematics in Africa, to the possibility of creating coexisting generative communities that do still use some of the materials from the extractive side but to create a circular model of use of materials, equipment, and enhancements that fit their cultural and other heritage. He has also provided us with a diagrammatic language to describe the material and design flow by analyzing different modes of production from the perspective of interconnected values and generative justice.

Does Law Wear Out?

The chapter by David Howarth, a lawyer, an ex-parliamentarian from the UK, on "Does Law Wear Out?" asks if the law is more like engineering than we tend to associate it with. In doing so, he argues that lawyers create contracts, wills, and statutory laws that are akin to engineered artifacts in satisfying a client's needs. He further claims that law as any engineered artifact also fails in specific ways including not serving the purpose that it was designed for, and thus can either need redesigning or the law itself must change via reinterpretation by the courts. Laws are shaped and modified both by changing social and political conditions as law is primarily about engineering society. Legal artifacts exhibit the same behavior as an engineered artifact, especially in term of failures. In making his case how laws may wear out, he chooses what is known as the bathtub model of failures from reliability engineering and the sawtooth model of failures in software to look at the life cycle of laws. Using the UK legal archives, he chose two corporate taxation laws by looking at the amendments to the laws as well as the mention of the two versions of the corporate taxation laws (1985, 2006) in the cases during 2007–2016 as surrogate measures for failures. In this preliminary analysis, it seems to indicate that failure of Laws follows in some form the bathtub and sawtooth model found in engineering. Howarth has opened a very important avenue to explore the relationship between engineering and law more thoroughly.

Dealing with the Future: General Considerations and the Case of Mobility

The chapter by Georges Amar, a French polymath engineer, author, and poet, "Dealing with the Future: General Considerations and the Case of "Mobility," opens the question of prospective as against prediction in engineering. The rise of artificial intelligence and other methods have given us the power to predict more and more precisely, but the future is not the past and is not only the basis of predicting the future but also to reopening the future and hence the course of the future itself. He makes the case that it is not easy to open this future. We may be able to predict the future consequences of climate change but that does not point to opening new possibilities that require going beyond prediction to prospective. Prospective requires that we understand the current paradigms thoroughly to ask how they would become obsolete and to question if they are sufficient for accommodating the new and nascent. The idea of the prospective is to create a new paradigm, an understanding with an enigmatic or even surprising "prospective proposition". This is not an artifact but a "conceptive prospective" to oppose the notion of predictive. The role of the prospective is not to reassure or create anxiety about the future but to open it up for new framing of the problem and concept themselves. Amar, with this basic conceptual frame, shows how one could reframe "mobility" as we have seen over our history as moving from one physical locale to another. He uses the example of the mobile phone that went beyond the connection of two people in a specific location to where the phone itself became mobile transforming the communicative and transactional mobility of a person beyond a location. He points to other concepts such as the "Walking Bus" that makes walking a collective act with stops and entry to the "bus" for school children's transport to school. The essay is a provocation to think prospectively in the conceptual space of defining a better future rather than predicting the future based on the past.

Designing the Future We Want

Yoram Reich's chapter on "Designing the Future We Want", claims that the current state of dialogues between different social groups is broken and that to design our future we need design education to be pervasive in society. He also makes the claim that a design perspective is critical to understand the basic interrelated questions of the "who, what, why and how" of every design problem. The answers to these questions in a socioeconomic context illustrates the range of dialogue and social inclusion that is part of the direction society has taken in its technological development. Reich supports Langdon Winner's claim that artifacts do have politics. He goes on to claim that the politics can be discerned through the institutional structures and exclusion in participation in the design of artifacts that are physical, policy, or institutional. The current model of design processes is biased toward those with power and money, who are designing the system for us in this environment of broken dialogues. In the analysis of the who, what, why, and how of a designed artifact we get to see the power relations and the distribution of benefits of such artifacts to the society at large. The chapter proposes a framework for understanding design along the axes of Problem Identification and Solution (P), Social Inclusion (S), and Institutional Structures (I), which interact with each other to reveal the scope and anomalies in their conception and execution in several case studies.

Engineering Design and Society

In his chapter on "Engineering Design and Society", Shyam Sunder establishes the dichotomy between engineering design as a practice rooted in natural sciences and natural objects with specific objectives to satisfy a human need, while social sciences are engaged in characterizing aggregate behavior from the individual. In spite of this dichotomy, he acknowledges the interaction between engineering and society while questioning the extent to which social science as practiced can contribute to engineering as the laws of human behavior are not generalizable nor made stable in space and time. Given these premises, he proceeds to explore the interaction between engineering design and society and the role social science plays in effectiveness of engineering and interventions. He uses six themes in his explorations: (1) Information and Design—accumulation of scientific information leads to new products and is closely tied to societal transformations; (2) Replication and Scale—replication of information is relatively cost free leading to scaling that was not possible before; (3) Choice Criteria—engineering efficiency as a choice is precise and accurate in its measurement, while social moral choices are fuzzier and are not easily quantified to make normative judgement; (4) People—as sentient beings with free will, their individual and aggregate behavior is not predictable as it continually evolves; (5) Dispersed Information—dispersion of information across people makes it harder for any centralized approach that can anticipate all possible trajectories of a product; and (6) Evolution—product evolves through incremental evolution mostly except an occasional disruption through a revolutionary product such as the Internet. In general, engineering sciences' relationship to practice is closer than social sciences' relationship with society as they differ in terms of existence of a clear objective in the first case and an emergence in the second case with unstable patterns and predictive powers in the aggregate. Through the exploration of the relationship between engineering and society, this paper arrives at some suggestions on the role of social sciences in engineering. The suggestion is to go beyond just studies of society to study prior technological interventions and their failures as a guide to a methodology for anticipating the effects of new technology on society with the hope for invariant patterns of behavior that transcend time and space. The case is made for the need for a new approach to social science in the context of engineering design and technological interventions that change the direction of the society.

Engineering the Engineers: Thinking about Design in Designing Thinking

The collection of chapters in this section addresses the role social sciences can play in the study of engineering and its practice from four perspectives. They are only representative of all the possibilities of use and embedding of social science in the study and improvement of science and engineering practice. The first chapter is a report from a separate workshop on macro- and microlevel studies of science practice in an effort toward understanding science of teams and the management of divergent and convergent processes in the engineering and physical sciences. The second chapter is on a micro-cognitive study of creativity in terms of the challenging of the well-accepted hypothesis that inspiration for creativity comes from concepts that are far from the problem at hand, but rather that it comes from concepts that are conceptually close. The third chapter is on an embedding of a social scientist in a science project and the sociologist's experience as an observer to record and care for the project. The final chapter in this collection is also a micro-study of cultural and emotional biases of creative collaboration to identify the conditions that trigger creative thought.

The Art of Research: A Divergent/Convergent Thinking Framework and Opportunities for Science-Based Approaches

The first chapter is a synthesis, by Glory Emmanuel-Aviñaa, Christian D. Schunn, and coauthors, of the findings of a workshop on the "The Art of Science: Opportunities for a Science Based Approach", held at Sandia National Laboratories. In this chapter, the authors use models of thinking that correspond to exploration and decision-making in engineering and physical sciences research as divergent and convergent thinking processes. The chapter goes on to examine the conditions under which the current "Art of Science" takes place and have identified individual, team, and institutional challenges associated with the successful execution of the divergent and convergent thinking processes. The primary argument in this chapter is that the time is ripe for interdisciplinary science that studies engineering and physical science research not just from an abstract social science perspective, from the perspective of being able someday to improve the practice of that research.

Do the Best Design Ideas (Really) Come from Conceptually Distant Sources of Inspiration?

The second chapter, by Chan, Dow, and Schunn, is a microlevel cognitive study of creativity to test the conceptual-leap hypothesis of creativity. The primary hypothesis tested in this chapter was to answer the claim that most creative solutions come from concepts from more distant fields versus from closer fields. The work is based on challenging prior studies whose methodologies may have not been statistically significant, or which suffered from different design contexts even within experiments. The results in the literature were mixed and supported both sides: that sources of inspiration sometimes came from far and sometimes from near. Here, instead, the authors used a more precise measure of conceptual distance, performing an experiment that involved text-based analysis of over a dozen design projects where a record of sources of inspiration were documented in OpenIdeo, a web-based platform. Based on this analysis, the chapter reports in detail the methods and measures used to prove that in their experiment the closeness in conceptual distance was the primary important trigger in inspiring creative solutions.

Integrating is Caring?! Or, Caring for Nanotechnlogy? Being an Integrated Social Scientist

The third chapter, by Ana Viseu, is on the incorporation of a sociologist of science in the Center for Nanotechnology as a policy decision and the results of this experience from a personal perspective set within the frame of social studies of science and the role of a sociologist. The main claim of the paper is that sociologists' role in observing and documenting the process for developing technology in society is often narrower than it should be. Sociologists, even when invited by engineers to participate, are not always welcomed in the actual evaluation of the social and ethical directions of the technology. The paper is a real example of the difficulties of incorporating social scientists in technological efforts, and thus potential limitations on the scope of their role in futuristic technology projects.

The Role of Emotion and Culture in the "Moment of Opening"— An Episode of Creative Collaboration

The fourth chapter, by Sonalkar, Mabogunje, and Leifer, is a micro-team-based study of creative collaboration by studying the emotional and cultural biases and interactions in engineering design teams. The goal of the research presented is to understand the occurrence of patterns of behavior that lead to creative collaborations by examining and analyzing video records of student design experiments. The primary thesis of this chapter is that there is what they term a "moment of opening", when the team seems to build on each other's ideas. This moment of opening is indicated by emotional states expressed by the participants, tempered by cultural patterns of expression. This is an interesting study of understanding group-based interaction in the triggering of generation of ideas to create new solution possibilities.

In Closing

This volume brings together most (though not all) of the papers given during the workshop. The volume does not, unfortunately, reproduce the real-time discussions following the papers, which were also interesting. Here, though, we mention just a few of the themes brought up during those discussions.

First, in the education of engineers, how might social science be incorporated formally in an engineering curriculum, and, so as not to dilute the physical science part of the curriculum, would this perforce increase the length of the bachelor's degree? The University of Virginia, and many others, for example, have begun to integrate the social aspects of engineering throughout their curricula. However, unanswered is the question of what formal model will be most useful (and acceptable) to the accrediting agencies. Also unanswered was the question of teaching methodology in mixed engineering and social science courses. A mixed teaching methodology which starts with a problem then elicits solution concepts through the lens of the problem might appeal to students better than the traditional "macho" model of teaching fundamentals first, but might dilute the rigor associated with the fundamentals. Also unanswered was the question of curricula at various levels: beyond undergraduate engineering, the reports from TU-Delft, ASU, University of Vermont, and CMU included Master's and Ph.D. programs with room for a deeper interplay between the scientific, technical, and political contexts of engineering problems and solutions.

Another question: what is the definition of the "common good." It was felt that the idea of the common good is quite weak in the minds of the engineers, however how to define the common good is itself not obvious. However, the need to bring the idea of common good in engineering education was felt to be necessary. A possible way to approach it was discussed by identifying individuals, who have contributed to the common good through their approach to engineering. The role of Steve Jobs, Frederick Taylor, and Frank and Lillian Gilbreth in terms of their social orientation were discussed. The need for distinction between those whose goal was primarily profits versus the common good was emphasized, as well as the need to identify role models who provide alternate conceptions of engineering for common good, and who could inspire qualitatively different kinds of innovations not just in products but also in governance and care of the environment.

The final conclusion of the workshop was enthusiasm for continuing this dialogue between engineering and the social sciences as a way to address complex problems facing society today. We intend to have the book available online as open access as next step to continue the dialogue, both online and offline.

Eswaran Subrahmanian
Toluwalogo Odumosu
Jeffrey Y. Tsao

References

Galbraith, J. K. (1998). *The affluent society*. Houghton Mifflin Harcourt.

Hausmann, R., Hidalgo, C. A., Bustos, S., Coscia, M., Simoes, A., & Yildirim, M. A. (2014). *The atlas of economic complexity: Mapping paths to prosperity*. MIT Press.

Layton, E. T. (1986). *The revolt of the engineers. Social responsibility and the American engineering profession*. Johns Hopkins University Press, Baltimore.

Ostrom, E. (2005). *Understanding institutional diversity*, Princeton University Press.

Perrow, C. (2016). Personal communication with Subrahmanian.

Pollio, V. (1914). *Vitruvius: The ten books on architecture*. Harvard University Press.

Roth, Alvin E. (2002). The economist as engineer: Game theory, experimentation, and computation as tools for design economics. *Econometrica, 70*(4): 1341–1378.

Sladovich, Hedy E. (Ed.). (1991). *Engineering as a social enterprise* (Vol. 26). National Academies Press.

Subrahmanian, E., Westerberg, A., Talukdar, S., Garrett, J., Jacobson A., Paredis, C., Amon, C., Herder, P., & Turk, A. (2003). Integrating social aspects and group work aspects in engineering design education, *IJEE*.

Varian, H. (2002). Economic scene: Avoiding pitfalls when shifting from science to engineering, business section, New York Times.

Veblen, T. (1963). *The engineers and the price system* (Vol. 31). Transaction Publishers.

Wauer, J., Mauersberger, K., & Moon, F. C. (2010). Fredrich Retenbacher (1809–1863). In M. Ciccarelli (Ed.), *Distinguished figures in Mechanisms and Machine Science*, Springer.

Chapter 1
Innovations in Energy-Climate Education: Integrating Engineering and Social Sciences to Strengthen Resilience

Jennie C. Stephens

1.1 Introduction

As the world confronts the challenges of climate change and the multiple negative societal implications of fossil fuel reliance, energy systems are in transition from predominantly fossil fuel-based infrastructures to more renewables-based (Brown et al. 2015; Princen 2015; McKibben 2016). Beyond technological changes, this energy transformation is associated with changing assumptions about energy generation and consumption with the rapid expansion of efficiency, solar, wind, and other renewable energies; these changes involve complex social dynamics which researchers have begun to explore from multiple perspectives (Berkhout et al. 2012; Hess 2013; Turnheim and Geels 2013; Fri and Savitz 2014; Stephens et al. 2015).

The pace of innovation at the energy-climate nexus is accelerating; however, educational priorities have not effectively evolved to prepare students for the rapidly changing energy-climate landscape. Change is occurring with advances in distributed energy resources including innovations in renewable energy and new approaches to managing electricity demand, which are linked to growing awareness of the climate change risks of fossil fuel-reliant energy systems (Brown et al. 2015; Princen et al. 2015).

Adapting education to prepare society for inevitable but unpredictable changes at the energy-climate nexus is a critical aspect of the energy transition that offers huge opportunities for innovation, diversification, and engagement. Many students recognize future job opportunities in the rapidly changing energy sector, but their educational path to prepare for those jobs remains unclear.

J. C. Stephens (✉)
School of Public Policy & Urban Affairs, Northeastern University,
360 Huntington Ave, 360C RP, Boston, MA 02115, USA
e-mail: j.stephens@northeastern.edu
URL: https://jenniecstephens.com

E. Subrahmanian et al. (eds.), *Engineering a Better Future*,
https://doi.org/10.1007/978-3-319-91134-2_1

The "climate-energy-education nexus" is a transdisciplinary space where academic researchers, professional scientists, policymakers, educators, and the general public come together to address real-world social, economic, educational, and environmental problems (Dahlberg 2001; Casillas and Kammen 2010; Dale et al. 2011; Criqui and Mima 2012). Popular journalist Thomas Friedman suggests that we have transitioned to an "Energy-Climate Era" where a positive future depends on renewable energy innovation and committed societal attention toward climate mitigation and disaster reduction (Friedman 2008).

When energy education is broadened beyond engineering—explicitly including social and cultural perspectives of energy system change—a more diverse set of students including women and underrepresented minorities are more likely to be interested in studying energy and entering the energy workforce. A gender imbalance in the energy sector workforce is widely apparent, although it is not well studied or documented (Pearl-Martinez and Stephens 2016). Integrating energy and climate education with engineering as well as social sciences has the potential to recruit a more diverse set of students in both engineering and in energy.

1.2 Advancing Energy and Climate Education in Higher Education

At the university level, students who want to prepare themselves for jobs in the energy sector or at the energy-climate nexus are uncertain what their major should be: should they focus on engineering, economics, policy, government, or environmental studies? To effectively prepare the future energy workforce, universities need to adapt to develop courses, curricula, majors, and minors that prepare students for jobs at the energy-climate nexus. University extension programs also offer the potential to expand awareness and engagement on energy issues between universities and local communities.

The need for interdisciplinary STEM-oriented energy education in higher education is growing as energy systems around the world transition from predominantly fossil fuel-based infrastructures to more sustainable renewables-based configurations. Energy education for sustainability (EEFS) is an emerging interdisciplinary field with growing student interest and strong career opportunities in STEM and beyond, but the field has limited cohesion or community. The National Council for Science and the Environment (NCSE)'s Council of Energy Research and Education Leaders (CEREL) is working to advance energy education in US colleges and universities. Recognizing that problem-based learning is central to STEM education, an emerging network of energy-climate educators are highlighting that the energy-climate nexus provides a particularly salient context to recruit and engage diverse students in active learning. The energy-climate nexus is a metafield that integrates diverse and interdisciplinary technical and nontechnical disciplines.

Expanding energy-climate education has transformative potential for understanding opportunities for empowering individuals, households, communities, and organizations to engage with the challenges and opportunities of energy system change and climate change. Educational institutions are responding slowly to the energy transition and climate change; graduates are insufficiently prepared to address the complex challenges associated with energy and climate change.

1.3 Beyond Conventional Engineering: Engaged Learning at the Climate-Energy Nexus

Conventional engineering education emphasizes the content to be learned and how it might be learned, including the classroom environments and delivery methods. This emphasis has been appropriate as the scope has been the education of engineers to ensure graduates have a solid foundation upon which to build their careers. However, as we broaden the scope of engineering education, we need to consider influences beyond classroom knowledge and skills, and their delivery (Walther and Radcliffe 2015). Increasingly educators are aware that behavioral, cognitive, and affective engagement (and ultimately, achievement) in the engineering major are influenced by the degree to which a student's engineering self-identity is well developed, autonomous (e.g., self-determined), and fits with the realistic features of potential careers they may aspire toward. Further, these constructs (self-identity and career expectations) are shaped by multiple psychosocial factors including individual differences in confidence and self-efficacy, social network factors (including the sense of belonging, connectivity, social support, social strain), motivational factors (e.g., possible selves, grit), and behavior (activity, learning). Student engagement has long been recognized as a critical factor in academic achievement. Engineering educators have studied and validated a wide variety of methods to improve student engagement in the classroom (active learning, inverted classrooms, do-then-learn, etc.). Bridging curricular (structured) and extracurricular (unstructured) engagement offers unique possibilities for making engineering education accessible to a greater diversity and number of students. Such activities create learning atmospheres of joy, trust, courage, openness, and connectedness in a uniquely consistent way.

Integrating the energy-climate nexus as a critical and salient context for expanding engineering education has potential to provide valuable experiences of engagement and a powerful enhanced sense of societal relevance to engineering.

1.4 Advancing K-12 Climate-Energy Education

The interconnected challenges of climate change and energy system change provide a valuable context for improving and advancing STEM education also at the K-12 level in the United States. The *climate-energy nexus* offers STEM teachers a unique opportunity to build students' future adaptive capacity and prepare students for a rapidly changing future in which climate-energy knowledge and experience will be increasingly valuable in the workforce.

STEM researchers and educators have identified a set of grand challenges facing society that have critical connections with the future of energy, food, water, health, and national security (AAAS 2007; Bybee 2010a, b; Nature 2015). Preparing students to address these challenges requires an integrated and transdisciplinary approach that was clearly articulated in a recent special issue of the journal *Nature* on "interdisciplinarity" (Nature 2015). There is growing awareness within the STEM education community that sustainability grand challenges are the most promising context for advancing STEM education and creating a shared vision of STEM education's goals and purposes (Chacko and Jennings 2008; President's Council of Advisors on Science and Technology 2010; U.S. Department of Education 2010; Judson 2014). For this reason, leading science educator Roger Bybee urges STEM educators to place the grand challenges at the center of their work, arguing they are central to STEM education's theory of action (Bybee 2010a, b). This theory of action is clearly evident in the new K-12 Next Generation Science Standards (NGSS) and companion framework (National Research Council 2012), both of which address the grand challenges, including climate change, in substantive ways.

Transforming STEM education around the grand challenges amplifies the convergence between STEM education, sustainability science, education for sustainable development, and environmental education (Clark 2007; Wals 2011; Wals et al. 2014). Research indicates that the majority of teachers are not trained to address sustainability issues and many lack conceptual and practical resources, including high-quality STEM partnerships, required to engage students with these issues (Potter 2009; Anderson 2012a, b). Currently, there is little research on climate change education in general and energy education in particular (Jennings et al. 2000; Jennings and Lund 2001; Thomas et al. 2008; Jennings 2009; Anderson 2010, 2012a, b; Council of Energy Research and Education Leaders 2015), and the STEM education community lacks a coherent framework for integrating the two.

STEM teachers in many parts of the country are also experiencing new opportunities for integrating STEM content and experiences into personalized learning plans (PLPs) for students. To take advantage of these opportunities, teachers will need new professional learning opportunities that integrate the concepts and practices of STEM education, climate-energy education, and personalized learning.

1.5 Energy Democracy for a Changing World

Embracing the potential for educational innovations at the energy-climate nexus is also essential to strengthening democracy. Increasingly over the past decade, renewable energy advocates, climate justice activists, and social and environmental justice activists have joined forces to organize around a call for *energy democracy*

(Burke and Stephens 2017). This call for energy democracy is strategic: democracy implies a broadly appealing agenda for greater inclusivity, equity, and influence among communities involved with a transformation in energy systems. The call is also pragmatic: a massive shift of technologies within the modern energy sector presents innumerable challenges as well as potential benefits. Greater democratic engagement would offer communities around the country and around the world stronger mechanisms to steer energy system changes and shape the development of a more renewable-based energy future.

As energy systems transform to more renewable-based, technologies, infrastructures, institutions, and cultural practices are shifting to accommodate new norms of energy production and consumption with multiple implications for community resilience. The emerging landscape of renewable power provides opportunities for local communities, individual, and households to control, own, participate in, and share benefits from the energy sector; however, renewable deployment does not necessarily result in enhanced community resilience. Energy democracy is a novel concept and emergent social movement that connects energy policy and social policy by rearticulating energy systems as distributed public works that distribute social benefits among local communities. This movement extends the social demands of energy systems beyond access, reliability, and affordability to include a broad suite of environmental, health, and economic benefits. By explicitly connecting policy issues that are generally dealt with independently, energy democracy framing provides a social, political, and cultural framework to assess community and climate resilience in energy system change.

The energy democracy framing is fundamentally political (Burke and Stephens 2018), which raises some challenges for educational institutions. Given the pervasive grip that fossil fuel industries and their financial and political allies command over contemporary political life, energy democracy activists seek to bring out into the public sphere the hidden infrastructures, privatized decisions, and distant consequences of modern energy systems. The instinct to politicize energy transition reflects an implicit understanding that the transition from fossil fuel dominant systems to those based on renewables offers an unprecedented yet potentially unrepeatable opportunity. Energy system changes have potential to reorder dominant corporate-controlled power structures. In more distributed, renewable-based energy systems, different decisions and investments will be made, different groups of actors will be politically repositioned, and material structures as well as enduring social and ecological patterns will be reestablished. The form of politics used to steer the energy transition will greatly influence the possibility for more democratic futures (Mitchell 2013).

In other words, if governed largely to preserve existing power relations, the renewable energy political economy may replicate existing dynamics of power, continuing to strengthen the powerful and weaken the marginalized (Duda 2015). Energy democracy sees in the energy transition an unavoidably political process as well as a key opportunity for advancing renewable energy and democracy together. In this way, energy democracy stands in sharp opposition to the strategy of promoting renewable energy by any means necessary (Sweeney 2014).

This political dimension of educational innovations in the energy-climate nexus provides an even greater justification for the importance of broadening interdisciplinary energy-climate education. When energy and climate change are taught in narrow disciplinary-confined ways, learners miss the opportunity for understanding complex systems and cascading impacts of networked systems.

1.6 Conclusions

Innovations in energy-climate education provide a most valuable and salient context within which to explore integration of engineering and social sciences. The silos that are so effectively maintained in our educational systems are hindering the capacity to prepare students for the rapidly changing energy-climate landscape (Stephens et al. 2008; Stephens and Graham 2010). We need to innovate our educational approach to adapt to the current energy and climate realities of human society (Jorgenson, Stephens et al. in preparation).

The emerging concept of energy democracy provides an innovative lens to explore the transformative social change potential of energy system change in an era of climate change. Social structures and policy processes that reinforce and perpetuate the legacy fossil fuel system are responding to the accelerating momentum moving toward more renewable-based societies. As the connections between energy resilience and climate resilience are becoming increasingly apparent, our educational institutions and systems have multiple opportunities to embrace, adapt, and learn from current and future changes and strengthen innovative integration of engineering and social sciences.

References

AAAS. (2007). *Grand challenges of sustainability science symposium at the american association for the advancement of science annual meeting*. San Francisco.

Anderson, A. (2010). *Combating climate change through quality education*. Brookings Global Economy and Development.

Anderson, A. (2012a). Climate change education for mitigation and adaptation. *Journal of Education for Sustainable Development, 6*(2), 191–206.

Anderson, A. (2012b). Climate change education for mitigation and adaptation. *Journal of Education for Sustainable Development, 6,* 191–206.

Berkhout, F., Marcotullio, P., & Hanaoka, T. (2012). Understanding energy transitions. *Sustainability Science, 7,* 109–111.

Brown, L., Larsen, J., Roney, J. M., & Adams, E. E. (2015). *The great transition: Shifting from Fossil fuels to solar and wind energy*. New York, Earth Policy Institute, W.W: Norton.

Burke, M., & Stephens, J. C. (2017). Energy democracy: Goal and policy instruments for sociotechnical transitions. *Energy Research and Social Change, 33*, 35–48.

Burke, M., & Stephens, J. C. (2018). Political power and renewable energy futures: A critical review. *Energy Research and Social Science, 35*, 78–93

Bybee, R. W. (2010a). Advancing STEM education: A 2020 vision. *Technology and Engineering Teacher, 70,* 30–35.

Bybee, R. W. (2010b). What is STEM education? *Science, 329*(5995), 996.

Casillas, C. E., & Kammen, D. M. (2010). The energy-poverty-climate nexus. *Science, 330,* 1181–1182.

Chacko, T., & Jennings, P. (2008). Issues in renewable energy education. *Australian Journal of Environmental Education, 24,* 67–73.

Clark, W. C. (2007). Sustainability science: A room of its own. *Proceedings of the National Academy of Sciences of the United States of America, 104,* 1737–1738.

Council of Energy Research and Education Leaders. (2015). *Council of energy research and education leaders*. Retrieved October, 2015, from http://www.ncseonline.org/program/Council-of-Energy-Research-%2526-Education-Leaders.

Criqui, P., & Mima, S. (2012). European climate-energy security nexus: A model based scenario analysis. *Energy Policy, 41,* 827–842.

Dahlberg, S. (2001). Using climate change as a teaching tool. *Canadian Journal of Environmental Education, 6,* 9–17.

Dale, V. H., Efroymson, R. A., & Kline, K. L. (2011). The land use-climate change-energy nexus. *Landscape Ecology, 26,* 755–773.

Duda, J. (2015). *Energy, Democracy, Community*. Retrieved June 27, 2016 from https://medium.com/@JohnDuda/energy-democracy-community-320660711cf4#jtxijr47s.

Fri, R. W., & Savitz, M. L. (2014). Rethinking energy innovation and social science. *Energy Research & Social Science, 1,* 183–187.

Friedman, T. L. (2008). *Hot, Flat and Crowded*. New York: Strauss & Giroux.

Hess, D. J. (2013). Transitions in energy systems: The mitigation-adaptation relationship. *Science as Culture, 22*(2), 144–150.

Jennings, P. (2009). New directions in renewable energy education. *Renewable Energy, 34*(2), 435–439.

Jennings, P., & Lund, C. (2001). Renewable energy education for sustainable development. *Renewable Energy, 22*(1), 113–118.

Jennings, P., Lund, C., & O'Mara, K. (2000). *New approaches to renewable energy education.*

Jorgenson, S. N., Stephens, J. C., & White, B. (in preparation). *Engaging environmental education with the climate-energy nexus.*

Judson, E. (2014). Effects of transferring to STEM-focused charter and magnet schools on student achievement. *The Journal of Educational Research, 107*(4), 255–266.

McKibben, B. (2016). *Why we need to keep 80 percent of fossil fuels in the ground. YES!* (February 15). http://www.yesmagazine.org/issues/life-after-oil/why-we-need-to-keep-80-percent-of-fossil-fuels-in-the-ground-20160215.

Mitchell, T. (2013). *Carbon democracy: Political power in the age of oil.* London and New York. Verso Books.

National Research Council. (2012). *A framework for K-12 science education: Practices, crosscutting concepts, and core ideas*. Washington: D.C., The National Academies Press.

Nature. (2015). Why interdisciplinary research matters. *Nature, 525*(7569), 305.

Pearl-Martinez, R., & Stephens, J. C. (2016). Toward a gender diverse workforce in the renewable energy transition. *Sustainability: Science, Practice & Policy, 12*, 1.

Potter, G. (2009). Environmental education for the 21st Century: Where do we go now? *Journal of Environmental Education, 41,* 22–33.

President's Council of Advisors on Science and Technology. (2010). *Prepare and inspire: K-12 education in science, technology, engineering, and math (STEM) for America's future*. Washington: DC, White House Office of Science and Technology Policy.

Princen, T., Manno, J. P., & Martin, P. L. (Eds.). (2015). *Ending the fossil fuel era*. Cambridge, MA USA, The MIT Press.

Stephens, J. C., & Graham, A. C. (2010). Toward an empirical research agenda for sustainability in higher education: Exploring the transition management framework. *Journal of Cleaner Production, 18,* 611–618.

Stephens, J. C., Hernandez, M. E., Roman, M., Graham, A. C., & Scholz, R. W. (2008). Higher education as a change agent for sustainability in different cultures and contexts. *International Journal of Sustainability in Higher Education, 9*(3), 317–338.

Stephens, J. C., Wilson, E. J., & Peterson, T. R. (2015). *Smart grid (r)evolution: Electric power struggles*. Cambridge University Press.

Sweeney, S. (2014). *Working toward energy democracy* (pp. 215–227). State of the World 2014, Springer.

Thomas, C., Jennings, P., & Lloyd, B. (2008). Issues in renewable energy education. *Australian Journal of Environmental Education, 24,* 67–73.

Turnheim, B., & Geels, F. W. (2013). The destabilization of existing regimes: Confronting a multi-dimensional framework with a case study of the British coal industry (1913–1967). *Research Policy, 42,* 1749–1767.

U.S. Department of Education. (2010). *Transforming American education: Learning powered by technology*. Washington, DC, U.S: Department of Education, Office of Educational Technology.

Wals, A. E. J. (2011). Learning our way to sustainability. *Journal of Education for Sustainable Development, 5,* 177–186.

Wals, A. E. J., Brody, M., Dillon, J., & Stevenson, R. B. (2014). Convergence between science and environmental education. *Science, 344,* 583–584.

Walther, J., & Radcliffe, D. F. (2015). The competence dilemma in engineering education: Moving beyond simple graduate attribute mapping. *Australasian Journal of Engineering Education, 13* (1), 41–51.

Chapter 2
Technology, Policy and Management: Co-evolving or Converging?

Margot Weijnen and Paulien Herder

2.1 Introduction

The TU Delft Faculty of Technology, Policy and Management (TPM), established in 1992 as the Faculty of Systems Engineering, Policy Analysis and Management, currently formulates its core business as *comprehensive* engineering: '*At the faculty of Technology, Policy and Management we combine insights from the engineering sciences with insights from the humanities and the social sciences. Our mission is to develop robust models and designs in order to solve the complex challenges of today's networked, urbanized knowledge society. Our three closely collaborating departments address these societal challenges each with a different perspective: systems, governance and values. The smart combination of these three perspectives is at the core of Comprehensive Engineering*'. The three departments: Engineering Systems, Multi-Actor Systems and Values, Technology and Innovation represent core expertise in technology and engineering, the social sciences and the humanities, respectively.

In TPM's evolution from systems engineering to comprehensive engineering, the Next Generation Infrastructures research programme (NGI) played a crucial role as a catalyst. Besides the aspect of substantial long-term research funding, enabling

---

Abstract for NSF workshop on "Engineering a Better Future: Engineering, Social Science and Innovation," April 15–16, 2016 at Carnegie Mellon, Pittsburgh.

M. Weijnen (✉) · P. Herder
Department of Engineering Systems and Services, Faculty of Technology, Policy and Management, Delft University of Technology, Delft, The Netherlands
e-mail: M.P.C.Weijnen@tudelft.nl

P. Herder
e-mail: p.m.herder@tudelft.nl

E. Subrahmanian et al. (eds.), *Engineering a Better Future*,
https://doi.org/10.1007/978-3-319-91134-2_2

TPM to develop critical mass in terms of PhD students, NGI forced TPM to substantiate its interdisciplinary systems approach. NGI framed infrastructures as complex adaptive socio-technical systems which need to safeguard a range of public values. The field of infrastructure systems turned out to be a rich source of inspiring research questions and a trigger to leap from relatively small-scale research projects within disciplinary knowledge units to more challenging interdisciplinary projects and programmes. This brought TPM to reflect on the nature of interdisciplinary research and how to make socio-technical interdisciplinarity operational in terms of research methods and tools, and in education.

2.2 Socio-Technical Systems

Infrastructure systems defy traditional systems engineering approaches. Most infrastructure systems in the Western world evolved over decades, even centuries. Although each and every link and node of an infrastructure system is designed in the best traditions of engineering design, the system as a whole was never designed as an integrated system. System integration itself is an evolutionary process that advanced as a result of the progressive interconnection of local systems, regional and/or national systems. The physical establishment of pan-European railway, highway, electricity and telecommunications networks preceded the political agreement on the European economic union. Historical research revealed that the train journey from Paris to Moscow took less time around 1900 than today.

Infrastructure systems are always in flux, so that the notion of optimality has different meanings at different temporal and spatial scales. At the basic level of the physical network, new links are being added (or removed) at different spatial scales, whether at the level of international connections or at the level of new connections to new residences at the capillary level. Optimality from the perspective of an individual user may not coincide with optimality for another user or with optimality at the level of the national or international network. While the topology of the overall network is changing, the hardware is generally designed to last for decades. During the lifespan of the hardware, economic conditions and societal preferences are changing, new technologies come to the fore and end users impose new demands on the quality and reliability of service.

In interaction with the provision of infrastructure bound services, social norms and standards for personal and domestic cleanliness, comfort, mobility and social interaction have changed dramatically. Infrastructures apparently influence society and the economy at a far broader and deeper level than just providing the functionality they were designed to deliver. In turn, new patterns of behaviour in society and the economy enabled or induced by infrastructure provision, and changing societal preferences, shape the continued development of infrastructure in terms of technologies as well as institutions.

What all this boils down to is that the social domain cannot be considered a context variable; in infrastructure systems, the social domain is an indistinguishable part of the system. The provision—or the absence—of infrastructure services affects the lives of all citizens. Infrastructure affects a range of competing public values, which need to be (re)balanced time and again. Moreover, in the neoliberal setting of most of today's infrastructure systems in the Western world, infrastructure development is the accumulated outcome of decisions made by a multitude of more or less autonomous actors, each with their own interests, values and means. In other words, infrastructures are not just exceptionally large technological systems: they are truly socio-technical systems. As many of society's grand challenges are directly and indirectly related to infrastructure provision, it is part of our academic responsibility to educate students in dealing with the complexity of these socio-technical systems.

2.3 Socio-Technical Systems in TPM's Education and Research

The Faculty of Technology, Policy and Management established in 1992 brought a radical innovation to the engineering education portfolio of TU Delft with the start of an interdisciplinary programme in Systems Engineering, Policy Analysis and Management (SEPAM), including a BSc and MSc programme. Until then, a selection of courses from the social sciences and the humanities had only been available as add-on electives to the traditional engineering education curricula. With SEPAM, and the TPM research programme emerging in parallel, TPM aimed to bridge the gap between engineering experts and the strategic decision makers in industry and public policy. It turned out to be the start of a rich and rewarding process of discovery. Initially, the best effort we were able to make was to present the students with two world views: one obeying the logic of science and engineering, and the other one a world of strategic agendas and multi-actor decision-making. How to confront and combine the social and engineering perspectives for better decision-making on real-world socio-technical systems was a case by case learning process.

Recognizing the need for a more systematic approach, we explicitly formulated the questions of understanding and governing infrastructures as complex adaptive socio-technical systems as core research challenges of NGI. It is then that we made the leap towards the development of hybrid analytical tools in which social and technical rationalities were integrated, with serious gaming (SG) and agent-based modelling (ABM) as prime examples. In SG, real-world players are the decision makers, whereas the system they govern is simulated to the extent of technical detail required to ensure the richness of behaviour that the players expect from the system, without which the credibility of the model and the players' engagement with the game would suffer. In ABM, social actors are simulated as software entities which are programmed, using state-of-the-art insights from behavioural sciences, to exhibit decision-making behaviour as observed in the real world, for example, in terms of risk aversiveness, and under simulated real-world conditions, for example, in terms of incomplete information. Over the past decade TPM built an extended ABM platform for simulation of energy, industry and e-mobility, which now allows us to study interactions, e.g. between electricity and CO_2 markets, between interconnected national electricity markets and between electricity and gas markets, between globally dispersed production units across borders, while, for example, varying exogenous variables such as market design variables, technological learning curves and regulatory regimes. We are currently developing ABMs that model policy processes and policymaking as an endogenous process in the simulation.

Both types of hybrid models are especially in demand for support of strategic decision-making. They allow for experimentation with the regulatory context of socio-technical systems, showing patterns of system behaviour that may emerge as a function of certain governance variables. However, both types of models—in fact, any hybrid model, cannot be validated in the strict sense of validation. Even if historical system development can be reproduced, that in itself is not a true validation of the model.

The lack of rigorous validation seems to mistakenly disqualify hybrid models for productive use in times where policy is supposed to be evidence-based. On the contrary: ABM and SG allow for a far richer exploration of possible infrastructure system behaviour in the future than traditional technical or economic or behavioural models, which miss out on either the social or the technical complexity of the system, and which fail to address the interactions between the social and the physical dimensions. In the practice of strategic decision-making, however, it is not a matter of choice between traditional and hybrid models. The pertinent question is how to select the most relevant combination of analytical, optimization and simulation models and how to make sense of their outcomes. Hybrid models (not only ABM and SG) can contribute to that synthesis.

In the SEPAM education programme, the 3-year BSc programme focuses on thorough problem analysis. The 2-year MSc programme focuses on synthesis. Over the course of these 5 years, complexity is gradually increased in three parallel strands:

(1) Technical specialization in one of four engineering systems domains:
    - Energy, Resources and Distributed Infrastructures,
    - Mobility, Transportation, Logistics and Supply Chains,
    - Water, Climate and Urbanizing Delta's, and
    - Cybersecurity, Innovations and (consequences of) Emerging Technologies in Cyberspace.

(2) Governance (including policy analysis, public management, economics, finance, ethics, etc.).
(3) Applications of methods and tools to actual socio-technical systems.

The BSc and MSc programmes were developed to seamlessly complement each other, ensuring a coherent interdisciplinary programme with a progressive build-up of social and technical complexity over the course of 5 years. The initial hesitation of some of the large industrial employers of engineering graduates to hire TPM graduates was soon overcome; TPM graduates easily find appropriate jobs in both the public and private sector, and their starting salaries are higher than for most of the traditional engineers. With the admission of undergraduates from traditional engineering programmes into the SEPAM MSc programme, starting in 2016, TPM is now confronted with the challenge of sensitizing engineering BSc's to issues of social complexity, and to ensure they acquire a sufficient knowledge base in social science to successfully complete the SEPAM MSc programme. On the basis of our experience with the Management of Technology and the Engineering and Policy Analysis MSc programmes, which admitted engineering undergraduates from the start, we know that it requires great mental agility for engineers to accept the social dimension as part of the system rather than as a context variable, and to embrace the social sciences in their own right, rather than employing superficial notions of the social sciences in an engineering framework.

2.4 Concluding Remarks

The intertwined histories of the TU Delft Faculty of Technology, Policy and Management and the national Next Generation Infrastructures programme have resulted in intensive cross-fertilization between engineering systems education and engineering systems research, between engineering and social science disciplines, between utility sectors, and between academia and practitioners. This unique co-evolutionary process has resulted in a number of undergraduate and graduate curricula and research programmes which distinguish themselves for their holistic approach to complex engineering systems as socio-technical systems. The hybrid models developed to substantiate TPM's interdisciplinary approach have had ample impact on industry and public policy, and TPM's graduates are firmly established now in strategic positions. These promising results do not guarantee future success. Especially in the institutionally fragmented multi-actor setting of infrastructure systems, the value orientations of stakeholders play an important role in their decision-making. Addressing the need for value-sensitive design is one of the challenges yet to be tackled in engineering systems education and research. We are tempted to speculate that rather than co-evolving, the engineering, governance and value dimensions of engineering systems are converging into a new science practice, the 'science' of comprehensive engineering, for which appropriate performance metrics still need to the defined.

Chapter 3
Reconnecting Engineering with the Social and Political Sphere

Jameson M. Wetmore

Many students choose to study engineering because they want to make the world a better place. They have skills in math and the sciences and they want to use those skills to create technologies that solve problems. They want to make people's lives easier and they want to find answers to society's most pressing dilemmas. The promise that engineering can address many of the world's "Grand Challenges" gives them a sense of direction and purpose.

All too often, however, as students progress through their undergraduate programs, this vision and purpose get clouded. A recent study has argued that engineering programs foster a "culture of disengagement" and that the public welfare concerns of students enrolled in them decrease over time (Cech 2014). There are likely many factors for this. The focus of the first two years of most engineering programs is on math and science courses and these are rarely linked to engineering practice, let alone social needs. Ethics courses are rare, and when they are taught they typically focus on microethics—local responsibilities—rather than the broader social questions covered in macroethics (Herkert 2005). And even design courses, which are supposed to help students think independently and connect different sets of expertise, rarely include significant attention to the social and political aspects of the problems being addressed. Students often enter engineering programs with broad social goals, but as they navigate through the details of a technical education their perspective and aspirations can narrow significantly.

I frequently see this firsthand when I teach engineering undergraduates. I usually find that the ethical dilemmas that are typically presented to engineering students distract them from the most important ethical questions. But occasionally I will ask them one. For instance, I have asked groups of students what they would do if their boss asked them to design a technology that they knew would put hundreds of people

---

J. M. Wetmore (✉)
School for the Future of Innovation in Society, Arizona State University, Tempe, AZ, USA
e-mail: Jameson.Wetmore@asu.edu

E. Subrahmanian et al. (eds.), *Engineering a Better Future*,
https://doi.org/10.1007/978-3-319-91134-2_3

out of work or whether they would design a deadly explosive device that could be concealed in a working mobile phone. Usually, the vast majority of students agree that they would. When I ask them why they offer two justifications: 1. "It's my job." and 2. "If I don't do it someone else will." Once when I was faced with a room full of young, bright, and talented budding engineers in their fourth year of undergrad making these arguments I tried to insult them to get them to see the implications of what they were saying: "Do you all simply think you are cogs in the wheel? Are there no ethics in your work? Are you simply mindless drones that do what you are told?" To my great sadness, most of them simply nodded "yes." By the time of graduation, many of these students had abandoned their hopes of changing the world and decided that their role in life was to solve the small technical problems that are assigned to them.

Technology is widely acknowledged to be at least part of the solution to many of the world's complex problems. The trick is that technologies have to be woven into society carefully and there must be accompanying social and political changes that complement them if real progress is to be achieved. Engineers will be needed to make this happen. At least one prominent sociologist has even noted that the most accomplished engineers are quite good at weaving the social, political, and technological together. Michel Callon contends that engineers "construct hypotheses and forms of argument that pull these participants into the field of sociological analysis. Whether they want to or not, they are transformed into sociologists, or what I call engineer-sociologists" (Callon 1987, p. 84). If technologies are going to be part of the solutions society needs, the "cogs in the wheel" we train must transition into "engineer-sociologists." Those with narrow technical expertise will need to develop a broader perspective to solve social challenges.

Thankfully, despite the culture of disengagement in many undergraduate programs, there are students who hold out hope that they can make a difference. They comply when they are told to focus on ever smaller technical details. But secretly they aspire to do something bigger and look for outlets for their interest.

For the past decade, I have been working to create opportunities for engineers such as these. Below I briefly outline two of them. These programs are designed to reconnect these engineers with the social and political contexts they work in and which they would like to improve.

3.1 Science Outside the Lab

The first of these interventions is Science Outside the Lab (SOtL), a program developed by the Consortium for Science, Policy & Outcomes (CSPO) to introduce graduate student scientists and engineers to the policy process (http://cspo.org/program/science-outside-the-lab/). SOtL is typically held in Washington, DC and brings together approximately 14 science and engineering doctoral students. Over the course of one or two weeks, the students meet dozens of people who work at the interface of engineering, policy, and society including policymakers, regulators, funders, lobbyists, lawyers, and museum professionals (Bernstein et al. 2016).

Most of the students that participate in the program start with the belief that policymakers in Washington, DC are either ignorant, corrupt, or a combination of the two. The program introduces them to incredibly intelligent policymakers, many of whom have science backgrounds, who have the difficult job of juggling the variety of values and pressures that go into making decisions that impact thousands if not millions of lives. Both anecdotal evidence and detailed evaluations have shown that the program goes a long way toward helping students understand the importance of the social context of engineering.

Students find that science and engineering knowledge is an important part of policymaking, but that it cannot dictate what should be done. The knowledge is only useful when combined with a robust understanding of the social and political factors and implications at play. By both understanding and recognizing the inherent worth of the values of different people and organizations, students learn how to play a more productive role in the policy process. As their respect for policy rises, they lessen their criticisms of it. And as they learn the place of engineering in the process, they are better able to do work that can directly inform policy.

The program was designed to help students become more effective at their jobs as scientists and engineers. And there is evidence that it has done that. Some participants, for instance, have reworked their dissertation research to inform deliberations over regulation that they learned about in the program and their findings were later cited in Federal rulemakings. Over time, however, students have used the program as a springboard to nontechnical jobs. Several graduates have gone on to take up positions in state and federal governments. By opening their eyes up to the complexities of policy, many students are drawn to crucially important problems that their technical tools cannot solve. Through this process, they slowly become a bridge by which different sets of expertise can be brought together.

3.2 Community Engagement Workshops

The second intervention is a response to the recent trend in engineering undergrad programs to offer some sort of experience whereby the students get a design challenge to solve a problem faced by people in the developing world. Unfortunately, many of these programs remain focused on technical details and the resulting designs offer little to no help to the local people who were supposed to benefit. It takes an especially talented "engineer-sociologist" to understand the nuances of a foreign culture, community, and place.

In an effort to address this problem and better prepare students to engage with communities, the "Equity, Equality, and Responsibility" subgroup of the Center for Nanotechnology in Society at ASU developed a two day "Community Engagement Workshop" that has been held five times in the US, Canada, and South Africa (Harsh et al. 2017). The major goal of the workshop is to give the participants an appreciation for the social complexities inherent in transferring technologies to developing areas and to imbue them with a sense of humility.

There are three major learning goals: 1. Decenter Technology. The students are shown the number of ways that a focus on specific technological artifacts can lead to enormous failures in development efforts and are encouraged to consider technologies as a possible component of a solution rather than the focus of a solution. 2. Listen to and learn from communities. The majority of development efforts fail because the outsider seeks to address a problem that the local people do not see as a problem. The course introduces students to ways to listen to a local community to better understand the community's concerns rather than making assumptions from the outside. And 3. Empower Communities. Many development projects work fine when the outsiders are still in the area, but as soon as they leave the programs fall into disrepair. One major step that can be taken to change this is to empower the local people to play a role in the project from the very beginning.

These workshops do not fully prepare students to have successful interventions in the developing world. Such a training program would take weeks, if not years. Rather, the exercises are designed to help the students begin to see the complexities involved in integrating any technology into the practices and institutions of a group of people. This brief introduction allows students to make a more informed decision as to whether they want to dedicate the time and effort necessary to make a direct positive impact in developing areas.

3.3 Conclusion

Not every student who participates in these programs will become proficient "engineer-sociologists." That is not a problem, however, because ultimately that is not the goal. We do not portend to teach the engineers involved everything they need to know about technology and society.

The graduates of these programs are, however, changed in two fundamental ways: First, often once they get an introduction to these ideas they want to know more and they seek to continue their education. They go on to study those issues that are most important in their field or for themselves personally. These programs often rekindle interests that have long lay dormant in thoughtful individuals and help them to find ways to reintegrate them into their lives and practice. Second, students come to better realize what they don't know. This acknowledgment of a deficit frequently leads them to better understand the importance of consulting other forms of expertise. An engineer who knows how to bring together the different experts necessary to solve important problems can be as beneficial as a lone "engineer-sociologist." Sometimes bridges are best made out of a chain of many people working together.

These interventions demonstrate that it is possible to break students out of the "culture of disengagement" that seems to be present in many engineering programs. Many of the students drawn to them knew that there was something missing in their education, but couldn't quite put their finger on it. We have seen the scales fall from the eyes of many of them. The experience often reinvigorates the students and they themselves are empowered to play key roles in productive change. There are ways to turn "cogs in the wheel" into leaders and bridge builders.

References

Bernstein, M. J., Reifschneider, K., Bennett, I., Wetmore, J. M. (2016). Science outside the lab: Helping graduate students in science and engineering understand the complexities of science policy. *Science and Engineering Ethics*, posted online 28 September.

Cech, Erin A. (2014). Culture of disengagement in engineering education? *Science, Technology and Human Values, 39*(1), 42–72.

Callon, M. (1987). Society in the making: The study of technology as a tool for sociological analysis. *The social Construction of Technological Systems: New Directions in the Sociology and History of Technology*, 83–103.

Harsh, M., Bernstein, M. J., Wetmore, J., Cozzens, S., Woodson, T., Castillo, R. (2017). Preparing engineers for the challenges of community engagement. *European Journal of Engineering Education, 42*(6), 1154–1173.

Herkert, J. R. (2005). Ways of thinking about and teaching ethical problem solving: Microethics and macroethics in engineering. *Science and Engineering Ethics, 11*(3), 373–385.

Jameson M. Wetmore is an Associate Professor in the School for the Future of Innovation in Society and Co-Director of the Center for Engagement & Training in Science & Society at Arizona State University. His work combines the fields of science and technology studies, ethics, and public policy to better understand both the interconnected relationships between technology and society and the forces that change those relationships over time. His recent work has been largely focused on two audiences. He works closely with scientists and engineers, often as part of the National Nanotechnology Coordinated Infrastructure, to build an understanding of the context of their research to increase their effectiveness in developing useful knowledge and technologies. And as part of the National Informal STEM Education Network, he collaborates with science center professionals across the country to help them embed discussions about the social aspects of present and future technologies on the museum floor. He is co-author of *Technology and Society: Building our Sociotechnical Future* (MIT Press, 2008).

Chapter 4
Ecole des Mines de Paris: A Few Lessons from a Long History

Armand Hatchuel

I am honored that the organizers of this forum[1] have asked me to present you with a short history of the Ecole des Mines de Paris.[2] I'll have to be brief, but I would like to explain how we went from a Vocational Training School to a General Education School, then to a School based on Research, and how the image of engineers has changed over the last two centuries. The status of engineers as developed in France in the middle of the nineteenth century was invariably that of a scientist, but a scientist who also accepted explicit responsibility for his acts. The responsibility was progressively defined by three functions—critical, creative, and social—whose content and relative priorities have changed and will continue to change with the major movements in techniques, sciences, and society.[3] For the Ecole des Mines de Paris,[4] we will see that these functions have been reinterpreted over time. I will present a few ideas on what this long history may mean for us in the face of our contemporary challenges.

To appear in: Proceedings of the International Dean's Forum, "Creating talents for a new world" August 2015 Paris. Presses des Mines, 2017.

[1]This forum commemorated the School's move to Boulevard Saint-Michel in 1815.
[2]I would like to point out that I became interested in the history of engineers, and more particularly the history of this School, as a teacher of management sciences and as the co-director with Benoit Weil of the "Design theory and methods for innovation" chair.
[3]Our work exclusively deals with the history of civil mining engineers. A history of the functions of the Corps of Mining Engineers comes from a very different approach, especially given that, since the middle of the twentieth century, they have received training separate from that given to civil engineers. We could point out that they share the scientific identity of engineer with civil engineers, this being reinforced by prior training at Ecole Polytechnique for most of them.
[4]The Ecole des Mines de Paris is now known as MinesParisTech.

A. Hatchuel (✉)
Chair of Design Theory and Methods for Innovation, MinesParisTech-PSL Research University, Paris, France
e-mail: armand.hatchuel@mines-paristech.fr

E. Subrahmanian et al. (eds.), *Engineering a Better Future*,
https://doi.org/10.1007/978-3-319-91134-2_4

A tradition of historical thinking

History has often contributed to collective thinking and orientations at the School. In 1889, at a time of major reforms, Louis Aguillon, Professor of Law, was asked by the board that ran the School to draw up a landmark historical report. The School was one hundred years old and the road traveled appeared to have led far from the initial project, giving rise to questions as to the choices made. This is where the historian's work takes on its full value because, as Aguillon pointed out, "*We can better explain the reason behind current things when we know what circumstances they were built upon.*" Since then, all the School's directors have supported historical thinking.[5] The same holds true for the teachers at the School.[6] But History cannot be studied without archives and without accessible documents, and the School has always benefited from a high-level library which has preserved an extraordinary wealth of materials—which is now available online.

As a reminder of some of the main chapters in the School's history, I will concentrate on how course content, pedagogical methods, the place of research, and views of "the engineer" have changed over time.[7] I am thus in line with the subject of this forum which invites us to think about what the talents of the future will be. But it is also through pedagogical questions and the paradigms of the engineer that the School's unique history can dialogue with the international history of engineers and their training. These questions enable us to situate cultural and national particularities, both in terms of the exchanges that the School has benefited from and the influence it has had. They shed light on the challenges shared by all major scientific training programs today.

To simplify things, I will make a distinction between three periods in the history of the Ecole des Mines de Paris. The first period was that of the *Professional Mining School*. This period came to an end around 1890. But the Professional Mining School did not disappear, it became part of a *Generalist School* that sought to train engineers adapted to all sectors of industry and more importantly their evolution.[8] At the end of the 1960s, this second School in turn became part of a *School based on Research*. This does not mean that scientific research only dates from that period, but rather that the School began to be organized around research centers designed to produce and transmit knowledge adapted to a world that is innovating faster than ever before.

[5]Please allow me to evoke the memory of Jacques Levy, who kindly encouraged my first research.

[6]I would notably point out the efforts made by our colleague Robert Mahl to diffuse this history.

[7]This text uses—in a different form—part of the material published in: A. Hatchuel, "La naissance de l'ingénieur généraliste: l'exemple de l'Ecole des Mines de Paris." Publication de l'Ecole des Mines 1997.

[8]In my 1997 study, ibid. *Hatchuel 1997*, I distinguish between two subsections in each of these periods. I spoke of a "widened professional model" between 1850 and 1900, and I made a distinction between a "sectoral general model" from 1900 to 1949 and a "universal general model" from 1949 to 1969. This text stopped in 1966, so I did not discuss the model of the School based on Research.

Each period developed a richer, more open concept of engineers. Their scientific identity has constantly been reasserted, but we have become more aware of the multiple functions that they have to be prepared for. This evolution does not follow a brilliant, omniscient pedagogical plan. It is the result of hard choices and inventions made necessary by the history of the world and sciences that have always held surprises over these past two centuries.

The starting point gave no hints of such an evolution. Around 1750, the noble science was that of the engineer-architect and specialists in military and civil construction. In 1729, Forest de Belidor's work, "*La Science des Ingénieurs*" (The Science of Engineers), became the reference and was quickly translated into German.

Faced with this noble science, which was already highly mathematical and whose exploits were visible—inherited by the Ecole des Ponts et Chaussées—the art of mining lacked prestige. It was mainly based on practice, trade, and observation. Working conditions in the mines were difficult and unhealthy. And yet, at the time, Saxony had a wealth of mining activities and had set up a New School, the Bergakademie, in Freiberg, which was famous for its teaching of the Art of Mining. Here, France copied Germany and took inspiration from the Freiberg School to found the Ecole Royale des Mines in 1783.

In 1815, with the fall of France's First Empire and following a number of ups and downs, the Ecole des Mines de Paris moved into its current home on Boulevard Saint-Michel. But the teaching was still the same as when it was created and reflected its mission, which was to be a Professional Mining School.

4.1 Professional Mining School: 1815–1890

The curriculum included four main courses, each spread out over 2 years. Three courses were scientific: *metallurgy, mineralogy and docimasy, and mineralogy and geology*, and one *Mine and Machine Operations* course which was to have a long-lasting posterity. This course is precisely where the image and role of mining inspectors and engineers were framed and discussed.

The aim was to teach an enlightened, responsible Art of Mining, as the first function of the students was *a critical function*: it was a question of fighting against inefficient, unjustified processes, bad practices by mine concession holders who exploited their mines poorly and endangered miners' lives.

But academic training alone is not enough for this mission. For a full mastery of the Art of Mining, solid, varied experience in the field is needed. That is why travel-study periods were so important, sometimes taking students to distant mining areas, and during which they kept a precise, accurate journal. These journals are now a major part of the School's great heritage.[9] What was the purpose of these trips? To discover the variety of mines and mining techniques from around the world, of course, but also to demonstrate, above and beyond their critical function, the students' real capacity to discover *their social function* in the difficult world of miners, which is not easily accessible and occupy the most foreign of regions. Thus, the School was right away involved in globalization.

The Professional School's mission appeared to be clear and well suited to France's domestic challenges. But new dilemmas appeared starting in 1850, requiring reflection.

First of all, there was a radical transformation in the art of mining itself, which led to a series of new basic sciences. Crystallography broke new ground in mineralogy. Thermodynamics, which is the science of transformations of energy and machines, shook up Bélidor's physics and required new mathematics. Chemistry, which is so important to the art of mining, underwent a revolution that changed all sciences. From then on, the traditional subjects taught at the School required a much higher general level in sciences, and this need was met by creating preparatory schools in mathematics, physics, and chemistry for the so-called "external" students, i.e., not from the Ecole Polytechnique.

Mining engineers continued to be people skilled in the Art, because each mine is unique. But they also shared the concept of the engineer, as an "applied scientist", which took hold during the same period. This identity contributed to engineers' especially high social standing in France. They benefited from an image as intellectuals with vast scientific culture compared with "the engineer" in the English-speaking countries, who maintained the image of a technician.[10] Furthermore, this new identity gave the engineer's critical function a more asserted legitimacy: alongside the subtleties of training in the Art of Mining, their scientific knowledge enabled them to make thorough analyses of, and sometimes to challenge, existing procedures.

During this period, the engineer's *creative function* became clearer than it had been before: the function of *industrial inventor*. This is explained in the major reference work of the day, Laboulaye's dictionary of arts and manufactures: "*The second degree of intellectual work is that of the engineer, the industrial inventor who, with a special purpose, applies acquired knowledge to industrial practices and every day creates new progress, new increased wealth.*"

This new concept oriented the School's subsequent choices, as events continued to arise that led the School, after long discussions, to add three courses that should be mentioned as they resolutely moved away from the initial professional model and foreshadowed the debates with which all engineering training programs were repeatedly to be confronted.

---

[9]For example, online you can read Edouard Sauvage's 1874 travel journal from his trip across America from East to West.

[10]This image is still very important in both the USA and the United Kingdom, with a multitude of campaigns designed to convince people that engineers are scientists.

New dilemmas and new courses

The first of these courses was *railways*. Should this have been considered a completely different industry and not teach the fundamentals of this activity? After all, the railways were replacing the ships and coaches that the School did not cover in its curriculum. But it is also true that the railways used techniques that came directly from the mining business. Lastly, it was an act of national interest and its wide scope was to make it the leading employer of engineering graduates. The decision to introduce this course was a wise one, as can be seen in the many descendants that were to follow. But was it the School's role to provide training for all new activities in the industry? Railways were a *generic* activity, i.e., necessary to all the others and an amazing source of new inventions. A different choice was made later on concerning the automobile and aviation.

The second course that broke with the past was *Law and Industrial Economy*, which is an emanation/specialization in the Mining School's operating course. The School was concerned when it was introduced—political economy was a sensitive, contentious subject. But the major developments in social questions needed to be dealt with. Child labor was banned in 1841. Students needed to be made aware of their new legal responsibilities, but without dogmatism and maintaining an engineering point of view. Aguillon gave a reminder of this: "*The council, considering that it is a question of applying political economy to mines and factories, was of the opinion that these lessons should be entrusted to an engineer who alone could understand the existing relations between mining laws and the questions of the art.*"

The third major course that gave rise to debates was *Paleontology*, which also lies on the edges of the professional model. During the first half of the nineteenth century, Natural History became a discipline that was inseparable from geological studies: Cuvier, Lamarck, and Lyell were famous, but here again there were many controversies, notably with the Church. There were careful attempts at a few conferences, then the idea of having a course was accepted, not without concerns: "*It was feared that the School would be diverted from its destination by producing naturalists rather than engineers.*"[11]

These three examples were forerunners to the now recurring dilemmas facing all engineering training—how should we react to the emergence of new industries? How should we integrate changes in law, social doctrines, and engineers' responsibilities? To what extent should scientific progress and the appearance of new sciences be taken into account?

In response to this, the School appears to have forged a guiding principle for its approach to novelties. This consists in not integrating new fields until it is possible to absorb their content, to adapt them to needs or even to develop them in a direction that is of interest to the School. Far from freezing the program in time, this principle was to organize the transition from the Professional School toward a new model that we will call the Generalist School, which preserved the content from the Professional School while inserting it into a more universal program.

---

[11]Ibid. Aguillon 1889.

4.2 Generalist School: 1890–1967

This transition had already begun to manifest itself in the 1870s when the effects of the second industrial revolution began to be felt, notably with the growing importance of industrial chemistry and the development of the science of electricity. Furthermore, the concept of the *laboratory* took hold as a teaching tool as well as progress in knowledge.

At the end of the 1880s, Henry le Chatelier developed the new "industrial chemistry" in his laboratory, which was his claim to fame. The opposite was true for electricity, for which there were just a few conferences and its main teaching was shoved off to the Machines course, which had replaced the mining operations course. At the same time, there was a trend among the main courses at the School to divide up into specializations: the geology course gave rise to a new branch with the Petrography course.

In 1885, the *Legislation and Industrial Economy* course separated into two courses. The new course was strongly influenced by the work of Frédéric Le Play, who invented social economy by developing his famous studies on working-class family life in Europe. This teaching was also linked to the development of the *encouragement society in social economy*, which had a strong influence on the industrials of the period.

At the end of the nineteenth century, the School appears to have been able to take a breath and contain the expansion of its subjects. Aguillon stated, "*They wanted to stay within the specialties that explain and justify the existence of the Ecoles de Mines, they did not give into the temptation to appear to be teaching everything, with the risk of teaching students nothing.*" In 1900, the School's director, Carnot, also insisted, "*The main goal of the reforms was to increase courses, in response to new branches in Industry... They avoided a general preparation that is necessarily insufficient for any career in industry and, conversely, they sought to delve as deeply as possible into all knowledge concerning the mineral industry....*"

But history did not come to a halt, and it even started to speed up. Since 1890, the automobile and aviation industries had become an ever greater presence. What should be done? Once again, a few specialized conferences were attempted, but they did not repeat what had been done with the railways: there were no grand, special courses. A course in nonferrous metallurgy was created, however, in response to the needs of new industries.

Moreover, alongside the old legal questions, *questions of industrial organization* were increasingly on the table. With Henry Le Chatelier, the School gave an enthusiastic welcome to Frederick Taylor and *Scientific Management*. The idea of "industrial science" was already very present at the School and its extension to questions of labor and organization corresponded to a new scientific approach to the engineer's social function, which took off in the 1950s.

From 1920 to 1949, countless industrial and scientific innovations came to the School. The number of teaching programs increased considerably. The School was de facto a generalist school, but it had not yet accepted this in its mission statement, and it did not yet have pedagogical structures adapted to the expansion of this model.

A new pedagogical model

That came about in 1949, when the School no longer hesitated to take a position that was the opposite to that of 1900: "*We can define the School's curriculum as a general technical curriculum.*" "*The extremely fast evolution and development of industry—notably the mining and metallurgy industries—make it no longer possible to hope to delve in depth into all the major techniques and sciences that students may need.*" Furthermore, in the field, "*engineers are confronted with much greater specialization than before….*"

A new pedagogical model therefore had to be formulated for the future: "*We will therefore prepare students for the range of specializations,… We will not sacrifice our training to the benefit of technology,… we will not shirk a difficult compromise, … and the focus will be on the work method,… and awareness of the mission of the future engineer.*"

This model gave rise to a new study structure which now had three levels:

(a) *General scientific education* including the classical sciences (math, physics, chemistry) as well as natural sciences in the mining tradition. This preserved the dual epistemology of modeling and observation.
(b) *Training in generic technologies* which, in principle, are found in all industries and which include law, economics, and scientific management.
(c) Lastly, *options*, which are specialized courses in which the School's educational traditions can be maintained: personalized tutoring, as well as teaching based on industrial internships and travel.

At the end of the 1960s, this model opened the door to new, original courses, but also to multiple options. It notably made it possible to imagine a new kind of engineer combining two mutually reinforcing identities:

- *A scientific identity* that is no longer just "applied" because they have to develop new sciences as well as new techniques and, for this, they must mobilize the most rigorous, most effective methods. These skills ensure their critical and creative functions.
- *An identity as a modern manager* because they have to take on their social function as organizers and managers in the business world based on the best scientific analyses of these questions.

This model corresponded relatively well to the needs of students after the events of May 1968. It also brought many young teachers to the School, who were to become the academic management of the new model of a *School based on Research* that absorbed the Generalist School, which was not sustainable over the long term without new resources and major institutional changes.

4.3 School Based on Research: 1967–2014[12]

We cannot talk about the institutional changes of the 70s and 80s without mentioning the influence of Pierre Laffitte.[13] But his influence would have quickly eroded if it had not been supported by the State and renewed by his successors at the head of the School.[14] What happened? Starting in the 1970s, the School undertook two interdependent changes:

(a) Implementation of a new academic model that included:

- The creation of its own doctoral program (1983) along with the development of many scientific masters degrees with universities. The curriculum followed by the students at the Ecole des Mines was equivalent to a Master of Science degree in the English-speaking countries. This program was truly successful as the School now trains nearly a hundred doctoral students each year.
- The creation of research and teaching centers in a wide variety of fields, including economics, management, and sociology.

(b) Definition of an original research strategy, called *directed research,* aimed at developing fundamental work based on industrial or societal problems.

This academic model consolidates collaboration between research and companies with the development of *Armines*, a private, not-for-profit operator in public research. The creation of the Sophia Antipolis technology park, for which we have Pierre Laffitte to thank, was also designed to encourage cooperation between science and development on a regional scale; the School played a major role.

With unwavering support from the State,[15] these new institutions encourage the dynamism and flexibility of the new model of the School based on Research. The School has thus been able to: (i) attract scientific staff who establish stronger, faster links between training and research; (ii) quickly develop new disciplines and new research subjects[16]; (iii) support the strong expansion of collaboration with industry which is now a major share of its budget.

---

[12]I adopted the date of 2014, not as the end of the School based on Research, but rather because, that year, MinesParistech became a founding member of the PSL Research University community and this choice would affect its future history.

[13]To whom the School paid a warm tribute at the event celebrating his 90th birthday on April 16, 2015.

[14]Jacques Levy, Benoit Legait and Romain Soubeyran.

[15]Under the authority of the Ecole des Mines, formerly the Conseil Général des Mines, today the Conseil Général de l'Economie, de l'Industrie, de l'Energie et des Techniques Avancées (CGEIET).

[16]The School played a major role in training for new disciplines: Geostatistics, Earth Sciences, Computer Science, Energetics, Scientific Management, Sociology of Innovation, Digital Economy, Design Theory, Logistics and Production Systems, Hazard Sciences and Quantitative Finance, etc.

This model also brings new representations of engineers. Alongside the image of a scientist-manager, the engineer's creative function is once again emphasized along with that of the *designer and innovator scientist*, and the *scientist-entrepreneur*. Thus, at the start of the 1990s, under the influence of Gilbert Frade, the School created the "entrepreneurial act", inviting all students to develop their ability to take initiatives through personal projects.

The School also introduced several research programs and original teaching in the fields of design, entrepreneurship, and innovation, some of which have become famous in France and abroad.

4.4 What Should We Take Away from This Long History? Contemporary Challenges and New Images of the Engineer

Many lessons can always be learned from such a long, event-filled history. First of all, I think the School has learned *values and principles* that have been implemented and are widely shared:

On the pedagogical level:

- Training for engineers is not just a matter of transfer of knowledge and also includes *an educational project*.
- Pedagogical methods must maintain *dual epistemology*: on the one hand, there is mathematical modeling, on the other, there is observation, experience, entrepreneurship, and travel.
- Training for engineers cannot be isolated, it must maintain *symbiotic relations* with other forms of education, with business sciences, economics, and society, but also with Art and Medical Schools.
- We need to increase awareness of the historical reality among our students: for two centuries, the concept of what an engineer is has evolved; it is still strongly marked by national cultures, but history also testifies to influences crossing from one culture to another.

On the academic level:

- Scientific research should be at *the heart* of training for engineers.
- It must be *flexible and robust* because the School needs to quickly develop new disciplines, but always based on solid scientific foundations to avoid academic and media *bubbles*.
- Research adapted to engineering training must be *collaborative* to a large extent, because engineers' mission is to deal with problems that arise in industry and in society, and also because fundamental discoveries sometimes arise from original studies of these questions. Furthermore, history has shown that, by cooperating

closely with companies, public authorities, and other university programs, the School has been able to *foresee* several major transformations and to prepare its students.

– *And tomorrow?*

Are these principles still useful in the face of the major challenges awaiting us? We think so. These challenges concern all universities and, as in the past, the School will profit from following the most interesting initiatives developed in the developed world and in emerging countries. Personally, I will insist on three of these challenges.

First challenge: the globalization in higher education is an invitation for the School to take its place in the "global village" of the university system. The School has chosen to do this by joining the PSL community.[17] All the PSL establishments place great importance on the link between training and research and this point is crucial, because the engineers we train must remain high-level scientists. Furthermore, the cycle between the creation of knowledge and the development of innovations which is of interest for engineer training will continue to shrink.[18]

Second challenge: the digital revolution is shaking up our own operations as much as those of our industrial and social partners. What teaching will we dispense tomorrow? How will companies operate tomorrow? We can think about this alone, but we are convinced that by maintaining strong collaborative research we will invent original forms of initial training and continuing education. As for many other revolutions over the past two centuries, the School will have to be a significant player in these transformations.

Third challenge: there is no denying that our times are marked by an accumulation of environmental and social crises and imbalances. The excesses of financial capitalism, the increase in social inequalities, the environmental and climate threats, the political chaos that reigns in many countries, etc. Over more than two centuries of existence, the School has been confronted with many difficult situations. All we need to do is to mention its date of birth—1783—to remember that, from the start, the School went through a particularly critical period in the history of the country and the world. We have every reason to believe that, as it has done in the past, it will have to participate in producing new models for training engineers. This effort will go through scientific, political, and institutional revolutions in which the School will have to play an inventive role. On the road forward, there are a few lessons from its long history that can serve as guides.

It is aware that the scientific training that engineers require entails close work with research. The School has also learned that the identity of its engineers is constituted through the three functions—*critical, creative, and social*—that we

[17]PSL Research University is a group of establishments set up under France's new laws and guidelines.

[18]MinesParisTech also has an association agreement with Institut Mines-Telecom, which falls under the same authority at the Ministry of the Economy: The Conseil Général de l'Economie, de l'Industrie, de l'Energie et des Techniques Avancées. Lastly, the School is a member of ParisTech.

mentioned above. But what content will these functions have in the future? Will others have to be recognized and will students have to be prepared for them? The importance of collective thinking on these points cannot be underestimated. On this subject, all I can do is to make a few suggestions based on work carried out at the School.[19]

(a) In a world dominated by the rules of financial capitalism, it seems to me that *the new critical function* of engineers lies in their ability to resist the primacy of short-term financial logic alone. They can legitimately embody the value of productive investment and the need for radical innovation, whether private or public. Faced with the contemporary fears raised by a science that appears to be too blind to its own consequences, they can also try to demonstrate that the scientific approach can be oriented toward collective progress.
(b) Since the beginning of the digital revolution, engineers have been increasingly called upon to act as scientists who innovate by designing new techniques and new usages, and, more often than in the past, to create businesses. *The new creative function* thus seems to me to lie in developing robust, sustainable innovation, and design methods that are up to the huge challenges of preserving the planet.
(c) Engineers' *new social function* is the product of these last two functions. It is thus a question of developing more participative and more inclusive forms of work, notably for the conduct and completion of complex, innovative projects. The task is all the more demanding in that these projects are usually carried out at globalized companies and will call for the intervention of many entities (subsidiaries, partners, associations, public authorities, etc.) in different countries and cultures. Some one hundred years ago, thanks to the scientific approach to organizations and to work, engineers contributed to an industry that is more attentive to its personnel. Today, they also have to think of ways to limit the abuses of globalized firms by improving the legal frameworks and standards that govern them. At the end of the nineteenth century, engineers played an essential role in the formation of modern companies. They must continue to pursue this mission today by participating in creating more responsible businesses, including in the poorest countries. Furthermore, as in the past, the School will have to respond to the humanistic aspirations of its students and society by adapting to their contemporary and future content.

These proposals have no pretention of being exhaustive. Moreover, new, heretofore unknown priorities are sure to arise. That is one of the most certain lessons of history. But along the lines that we have discussed, the Ecole des Mines in 2015 is on the right road.

---

[19]Notably at the Economics, Business and Society Department. And more particularly, concerning my own research, at the Scientific Management Center.

Chapter 5
Evolving from Single Disciplines to Renaissance Teams

Dan Siewiorek

5.1 Origins and Motivation

In 1989, Carnegie Mellon University's National Science Foundation (NSF) Engineering Design Research Center (EDRC) brought in-house a stereolithography apparatus (SLA), providing three-dimensional additive manufacturing capability. As an experiment in collaborative multidisciplinary design, we created an SLA housing for a small single board computer that had been automatically generated by an electronic computer-aided design system (ECAD) that synthesizes computer systems from specifications (Gupta et al. 1993). The housing was designed by an electrical engineer (EE) and most resembled a Howe Truss bridge—the mechanical design taught to EE students at the time. In addition, the only way to reach the computer reset button was by thrusting a finger through the cooling fan blade. Thus was motivated the formation of our first multidisciplinary team composed of a software, electrical, and mechanical engineers, and a designer from fine arts to design and build our first wearable computer in 1991. VuMan 1 (Siewiorek and Smailagic 1993) had a shoulder strap supporting a housing with a raised hand rest allowing easy access to buttons for control. The hand rest also served as a chimney for convection cooling from the largest heat producing electronic chips that were clustered under the hand rest.

In the same time frame, Randy Pausch, co-founder of the Entertainment Technology Center (ETC) taught a course on Building Virtual Worlds (BVW) that engaged students from computer science, engineering, social science, fine arts, and design. Five student multidisciplinary design teams were formed that researched themes, wrote scripts, created choreographies, generated graphics, and animated a unique virtual world in only 2 weeks. Then the team membership was scrambled

D. Siewiorek (✉)
Human Computer Interaction Institute, Carnegie Mellon University,
Pittsburgh, PA, USA
e-mail: dps@cs.cmu.edu

E. Subrahmanian et al. (eds.), *Engineering a Better Future*,
https://doi.org/10.1007/978-3-319-91134-2_5

and another virtual world was created. The 2-week process was repeated five times in a 15-week semester. No one person could be proficient in the wide diversity of skills required to build a compelling virtual world.

Randy liked to point out to his students that they possessed more science and mathematical knowledge than Leonardo Da Vinci—the model Renaissance Man. But science and technology have advanced so much that no one person could possess all the knowledge to build contemporary systems. Thus students needed to become members of a Renaissance Team learning the vocabulary and appreciating the skills of other disciplines.

5.2 A Multidisciplinary Design Course

From this experience, we developed and employed a User-Centered Interdisciplinary Concurrent System Design Methodology (UICSM) that takes teams of electrical engineers, mechanical engineers, computer scientists, industrial designers, and human–computer interaction students working with an end user to generate a completely functional prototype system during a 4-month long course. Over 25 years of use, the last 20 years in a formal class setting, the methodology has proven robust in creating an increasingly capable applications leveraging state of the art components.

UICSM is web-supported and defines intermediary products that document the evolution of the design. These products are posted on the web so that even remote designers and end users can participate in the design activities. The design methodology proceeds through three phases: conceptual design, detailed design, and implementation. End users critique the design at each phase. Our experience has been that end users often cannot articulate what they want but they are able to critique concepts. By iteratively designing we can elicit functional requirements. Prior to each phase, students are given examples of the work products for that phase. Working in their subgroups they develop their work products and share them during class periods.

During the first phase—Conceptual Design—students work in discipline-specific teams. The Human Computer Interaction and social science students work with the end users to define baseline personas and "day in the life of" activities. Visionary scenarios are also created to explore how technology could improve current practice. The group develops functional requirements and an interaction architecture. In parallel, the technology students research available technologies: sensors, hardware platforms, and software development environments culminating in hardware and software architectures. During the second phase—Detailed Design—functional subsystems are identified and multidisciplinary teams are assigned to design and implement each subsystem. Each team is responsible for ordering components and designing their subsystem to integrate with all the other subsystems. During the final phase—Implementation—the subsystems are finished and integrated into a complete single system.

Figure 5.1 is an example of how the disciplines interact to achieve constraints on user attention, user interaction, manipulation, corporal, and power when designing a portable electronic system. The major design parameters are the functionality, user interface, physical form factor, and power. Electronic and software designers create the functionality taking into account of the division of **Attention** between the physical and virtual world. Software and Industrial designers define the **User** experience. Industrial Design and Mechanical Engineering define the **Manipulation** (e.g., controls quick to find and easy to operate) and **Corporal** (e.g., interface physically without discomfort or distraction) experiences. Electronics (heat generated) and Mechanical Engineering (dissipation of heat) interact on **Power**.

The methodology has been used in designing over three dozen computer systems, with diverse applications including inspection and maintenance of heavy transportation vehicles; augmented reality in manufacturing and plant operations; car/driver interaction; bridge inspection; aircraft sheet metal inspection; offshore oil platform crane operation; and welding nuclear submarine hulls (Siewiorek et al. 1994; Smailagic and Siewiorek 1999; Smailagic and Siewiorek 2002a, b; Siewiorek et al. 2001).

Communications between groups are essential for UICSM to be successful. Kiva combines aspects of both email and bulletin boards to keep threaded discussion intact. Visualization tools allow tracking of group progress and signal areas for instructor attention. We have created analytical methods based on Communication networks to study the content of the communication Isolated groups can be easily identified.

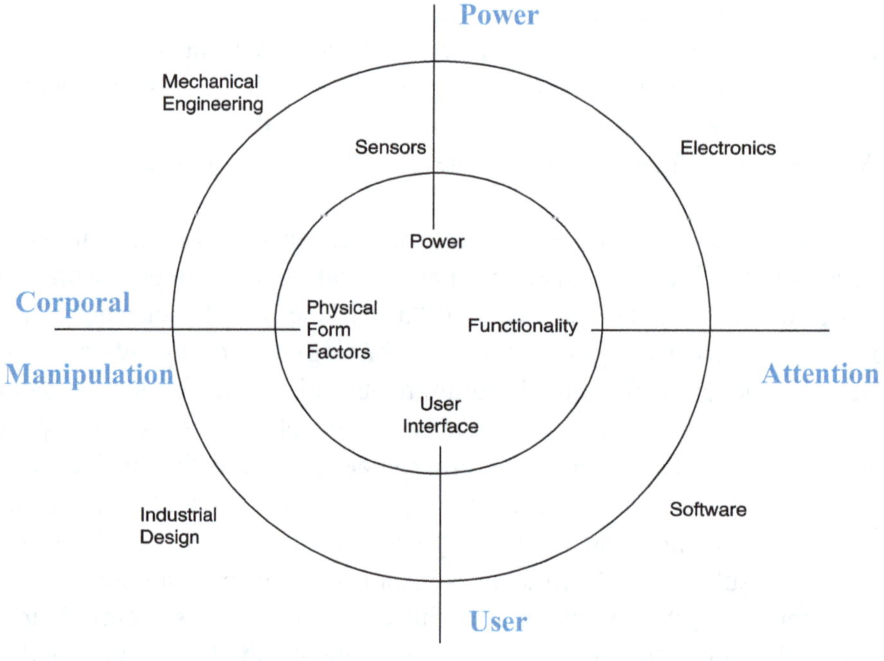

Fig. 5.1 Relationships between disciplines

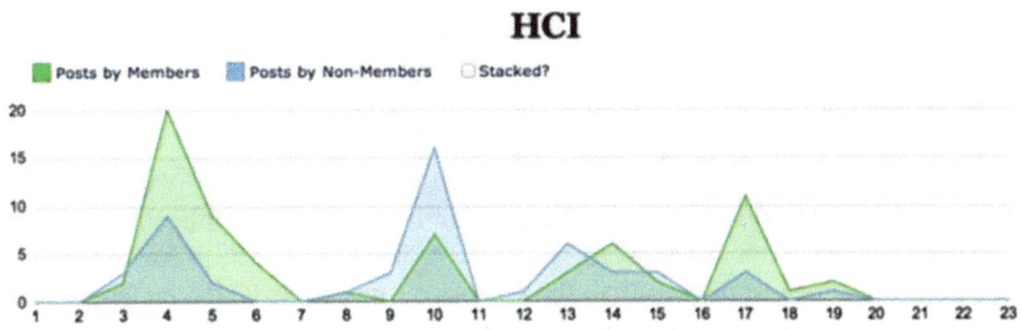

Fig. 5.2 New noun phrases introduced through three phases of design

For example noun phrases for analyzing how the design vocabulary of student teams expands and contracts during a design project. The expansion and contraction reflect the students brainstorming about different design alternatives for the project and finalizing certain design solutions, respectively. Figure 5.2 depicts the new noun phrases introduced through the phases of UICSM: conceptual design, detailed design, and implementation and integration. Reduction in the number of new noun phrases indicates a convergence in the phase and that the students have established a common vocabulary and vision of the design.

5.3 Lessons Learned

The User-Centered Interdisciplinary Concurrent System Design Methodology (UICSM) is used in a capstone engineering design course with 25–35 students representing 5000–7000 engineering hours for each system.

- The challenge for all the systems was designing an information flow that could utilize then current wireless capability and fit into the existing workflow.
- Several of the systems continued evolution beyond the course: in system maintenance a spin-off company developed Interactive Electronic Technical Manuals acquired by Boeing; formed the specifications for an all-new electric drive boat; formed the concepts for an all-new maintenance approach for Digital Equipment Corporation; and created the framework for tools and analysis supporting UICSM courses.
- Students apply their disciplinary skills in appropriate portions of the design process learning from each other.
- Students often ask what our expectations are for the system. Our response is that we do not want to bound their creativity. We give them example functions to stimulate "out of the box" thinking. The resulting systems always exceed our expectations.
- All the systems led to successful demonstrations through risk management and functional redefinition. We typically achieve about 80% of the visionary

scenario. While the students may feel disappointed, we point out that their system is a major advance over the baseline scenario as reflected by the enthusiastic acceptance by the clients.
- Several of the designs won international design competitions. One of the judging juries stated the design looked good enough to actually work, missing the point that the system was designed to fulfill a real-world application and had been implemented and deployed.
- Perhaps the greatest testimony to the success of the course is that undergraduate engineering students are filling out Institutional Review Board (IRB) protocols realizing that their designs need to be evaluated in the context of human use.

References

Gupta, A. P., Birmingham, W. P., & Siewiorek, D. P. (1993). Automating the design of computer systems. *IEEE Transactions on Computer-Aided Design of Integrated Circuits and Systems, 12* (4), 473–487.

Siewiorek, D. P., & Smailagic, A. (1993). A case study in embedded-system design: The VuMan 2 wearable computer. *Special Co-Design Issue of IEEE Design and Test of Computers, 10*(3), 56–67.

Siewiorek, D., Smailagic, A., Lee, J. C., & Tabatabai, A. R. A. (1994). An interdisciplinary concurrent design methodology as applied to the navigator wearable computer system. *Journal of Computer and Software Engineering, 2*(3), 259–292.

Smailagic, A., & Siewiorek, D. (1999). User-centered interdisciplinary design of wearable computers. *ACM Mobile Computing and Communications Review, 3*(3), 43–52.

Smailagic, A., & Siewiorek, D. (2002). Application design for wearable and context-aware and wearable computers. *IEEE Pervasive Computing, 1*(4), 20–29.

Siewiorek, D., & Smailagic, A. (2002). User-centered interdisciplinary design of wearable computers. In J. Jacko & A. Sears (Eds.), *The human-computer interaction handbook* (pp. 635–655). New Jersey: Lawrence Erlbaum Publishers.

Siewiorek, D., Smailagic, A., & Salber, D. (2001). Rapid prototyping of computer systems: Experiences and lessons. In *Proceedings of 12th IEEE International Workshop on Rapid System Prototyping*, Monterey Beach Resort, CA, June 2001.

Chapter 6
Designing the Future We Want

Yoram Reich

Design, if it is to be ecologically responsible and socially responsive, must be revolutionary and radical (going back to the roots) in the truest sense.
Victor Papanek, *Design for the Real World* (1971, p. 343).

6.1 The Dialogues Are Broken

Are we happy about the situation around us? U.S. elections and their aftermath, Brexit, and elections coverage elsewhere in the world tell us that many people are looking for a change, and it is unclear that they or most others understand exactly the consequences of these changes. There is fear of terrorism and security problems; there are cultural clashes, old and new; and the old problems of hunger and inequality, and controversies on global warming still persist. Yet, some technology advocates tell us, and they are more insistent today, that technology has the solutions to all human challenges, that with technology, the future is better.[1] Abundance is better, much as economic growth has been a goal in itself for decades.

Some problems indeed are addressed by technology but others emerge. This cannot be a surprise as we already know from the behavior of complex systems that sometimes, even a seemingly trivial combination could lead to results that no one

The title has two meanings. The first meaning is as it is read: a call for us to design our future as we want it to be. The second meaning claims that the situation of designing is the future we want. Further, this title embeds the title of J. Fresco's book, Designing the Future, The Venus Project, 2007, where he outlines his technological solutions to challenges and inviting others to participate in the design of their future.

[1]Diamandis and Kotler (2012).

Y. Reich (✉)
School of Mechanical Engineering, Tel Aviv University, Tel Aviv, Israel
e-mail: yoram@eng.tau.ac.il

E. Subrahmanian et al. (eds.), *Engineering a Better Future*,
https://doi.org/10.1007/978-3-319-91134-2_6

could forecast.[2] Consequently, such emergent behavior should be expected. When it has harmful effect, sometimes, it may be contained, but only if we detect it soon enough. Fast pace changes may be difficult or impossible to control and unfortunately, technology advocates tell us that progress is exponential.

The problem may also be in the concept of abundance we seem to strive for. Wu[3] makes a compelling case against it from several perspectives. First, it leads to problems such as obesity and information overload, where the latter may counter the positive effects of the wealth of information and networks.[4] Second, this wealth leads people to lose their ability to exercise sensible judgment.[5] While scarcity makes people agile, fit, and even creative,[6] abundance may make them unfit. It becomes easy to use artifacts for serving political purposes or minority benefits rather than dealing with general humans' betterment and the care of the environment.[7] Metaphorically, abundance seeks to get us back to the Garden of Eden but with it, perhaps undo the consequences of eating the apple: have whatever is offered to us and be dumb. Is this what we want?

It would have been great if we had clear knowledge about the consequence of our actions or what wrong could be done with the technology we develop. But as presented above, for each book that provides a compelling view with evidence about a situation, there may be another with a contradicting view. These views present models of the world, approximations that may reflect everything from reality to illusions.[8] They are well-argued opinions, but not truth. Shouldn't these opinions engage in a dialogue to sort their differences?

We saw that technology solutions may not lead to benefiting all stakeholders and are not even considering all stakeholders desires initially. They also do not necessarily lead to democratizing decisions. This has been noticed years ago, and similarly, the case for broken dialogues and their consequences have been made before. There have been proposals to address this problem toward a more humanistic approach to social change through the design of our environment, our lives, and the way we govern ourselves.[9] These clarion calls while celebrated at times have fallen on deaf ears as they have not had their effect on our daily lives.

---

[2]Wolfram (2002), Perrow (1984).

[3]Wu (2013).

[4]Benkler (2006).

[5]Tierney and Baumeister (2011).

[6]A great example of creativity and skills turning barrels into stoves https://www.youtube.com/watch?v=JBOF0OTPUrs.

[7]Winner (1980), Gilens and Page (2014).

[8]Such models can be created intentionally, by design, to serve the opinion proposed. It would be justified as following the scientific method where, for example, physical phenomena are removed from their context to be studied in the lab. However, in our case, the issues we discuss cannot be isolated from their context.

[9]Fuller (1960), Papanek (1971).

When dialogues are broken, other means are been used to resolve conflicts. These means have become more vitriolic, rhetorical, and violent resulting in civil conflicts and wars.

The question for all interested in fixing the dialogue is whether such a fix is a lost cause or should we develop new models for the creation of healthy and productive dialogues. To understand the task ahead of us we need to understand why earlier ways of addressing this problem have failed? Why is it that we cannot mobilize the people to support such ideas? Was it because they challenge the political order? Is the public given "bread and circuses"[10] to divert its attention? Are we doomed to wait for a catastrophic dysfunction of the world with no hope for a change? Do we really have no free will?[11] How do we approach our goal of change given that there are revolutionary efforts that do come about and flourish in some corners of the world?

6.2 The Design Hypothesis

Everything around us other than nature has been designed, contemporary human problems alike. One may claim, following the complex system argument (see Footnote 3), that humanity's challenges are accidents, emergent behavior out of past, and present unmanageable context. But given that they are not addressed seriously today, and that some people benefit from them, we can postulate that those who benefit exercise design to maintain the status quo. The design hypothesis contends that if there is a problem that persists, we should look for those who benefit from it and we will find out who designed the problem or is presently engaged in preventing its solution by further design. We exercise significant intention in our designing; otherwise, it would be quite difficult to explain huge coordinated efforts such as the Apollo Mission, the development of the F35 stealth fighter or the Large Hadron Collider at CERN. Until refuted, the pervasiveness of design as an activity that underlies the human condition is my working hypothesis.

With this hypothesis suggesting that humanity's problems or their maintenance are driven by design, it becomes imperative to better understand the nature of this critical human skill—designing. As a first observation, if we argue that there are critical challenges facing humanity, it means that the act of design has not been conceptualized and exercised well by most of us. If we see around us inequality, it means that some people exercise their ability to design their future very well to their advantage compared to others. This behavior is not imaginary as has been noted in

---

[10]Satire X by Juvenal.

[11]Sapolsky (2017). Sapolsky makes a case that our behavior is determined by our biology and that the reason we are unable to predict or explain fully certain behaviors is our lack of knowledge of the multifactorial relation between biology and behavior. This view is in complete contradiction to the design view proposed here but its present status is just a hypothesis.

recent study that shows how the American upper middle class has managed to create space that more and more ensures their social and economic mobility at the cost of those who are in the lower economic classes.[12]

It is not surprising that we would find diversity in designing skills; there will always be a distribution regarding any skill. However, the point is different; designing is a particular skill that is ignored in our educational systems, leading to destroying whatever design skills we are born with or acquire until kindergarten.[13] This deficiency creates a situation that most people lack design skills, creating an opportunity for others who maintain them or acquire them to be far ahead of others.

The first step is, therefore, educational, at all fronts, starting with young children; continuing to adults, and even professionals and researchers; and ending with the elderly. All these people face challenges or issues on a daily basis that require designing; they will all benefit from owning better design skills. To enlist them in an educational program, we need to improve their awareness of this critical human skill and the significant consequences of exercising it to our lives. Designing this educational project is outside the scope of this essay; its scope is explored elsewhere.[14] Here, I only describe briefly the status of the broken dialogue between disciplines and some attempts to fix it. I end with a framework about design called the PSI matrix[15] that helps put different disciplines and their interactions in perspective so that we can better understand what designing is, when it succeeds or fail, and subsequently, be able to design a better future.

6.3 Broken Multidisciplinary Dialogues

Let us go deeper to understand some causes of the broken dialogue and its extent. This could help us understand how to approach fixing it. Disciplines as we recognize them today have their roots in their formal study at higher education institutions. Take, for example, engineering that started as military engineering. In the nineteenth century, civil engineering and mining branched out as separate disciplines and toward the end of the nineteenth century, new disciplines emerged rapidly, chemical engineering, mechanical engineering, electrical engineering, material, agricultural, industrial, and others.[16] Each discipline started to concentrate on its own products, developed concepts, and models to deal with them, effectively specializing a language to converse among its members. It became difficult for professionals from different disciplines to effectively communicate.

---

[12]Reeves (2017).

[13]Refer to Ken Robinson' TED lecture, Do Schools Kill Creativity? for a popular articulation of this subject https://www.ted.com/talks/ken_robinson_says_schools_kill_creativity.

[14]E. Subrahmanian, Y. Reich, S. Krishnan, *We The Designers*, submitted.

[15]Reich and Subrahmanian (2015, 2017).

[16]Tadmor (2006).

As long as we stay within engineering disciplines, a fraction of these communication difficulties could be bridged by new theoretical developments,[17] simple visual languages that could form a common ground,[18] or lead to some advances in engineering education.[19] But the problem is much more profound. Throughout the twentieth century, engineering education in the different disciplines turned from stressing the practice or skills of the professions though forms of apprenticeship, to focus on theory and became engineering sciences. Engineering schools started to create people who knew the science in one niche area but little about the consequences, practical, technical, environmental, social or otherwise, of practicing their knowledge in the real world. From the master builders of the past, professionals became masters of siloed knowledge. This created professional engineers knowledgeable about some of the technical aspects needed for addressing challenges but often not even the ability to integrate knowledge within their own discipline and definitely ignorant of actual human needs and context. In a discipline such as economics, such educational practices led to people that are experts in the science of economics but do not know whether it is related to practice or real economics and may be uninterested in its consequences to issues such as sustainability. This situation, where there is no training in the synthesis of knowledge from different spheres, allows those in power to manage the present situation and even increase existing gaps such as between rich and poor.

Consider, for example, the development of technologies such as smart clothing or augmented reality. While they could be useful in many cases, they would constantly collect information about us in an unprecedented way. For example, there would be a timed record of where we look and how our heart beats in response. We could then be presented with filtered, summarized, and even manipulated information[20] that would change how we feel,[21] inform us about what we see, who we

[17]Such as the Interdisciplinary Engineering Knowledge Genome (IEKG) that attempts to create a bridge between different disciplines through a network of mathematical representations, Reich and Shai (2012).

[18]Such as visual languages in Quality Function Deployment that could allow project managers and systems engineers as well as production line workers to converse and exchange vital knowledge, Akao (1990).

[19]Crawley et al. (2007).

[20]Consider the recent ruling of the European Community against Google for manipulating its search engine results to gain advantage over its competitors. Google was fined for 2.42 B€. https://youtu.be/9cYeNpETjyw. The ability to do so through information accumulation and its further discussed in Zuboff (2015). Another way to distort information for negative purposes is available by social bots: Ferrara et al. (2016).

[21]See, for example, Kramer et al. (2014). Note the significant controversy about this study concerning its unethical status, not informing participants about their participation in the experiment, and the avoidance of Cornell University and PNAS journal editors to take responsibility for not checking it and publishing it in this way. Note also that the effect reported in this study is almost negligible but the sample size is so big that even such slim difference ends up being significant. Finally, since the first author is a Facebook employee comments on the web suggested that this paper will now allow Facebook to increase in advertisement fees given the impact that now could be associated with its information.

should be weary of, and what we must buy on our way. While we are so occupied with this stream of information, catastrophic accidents may occur.[22] Since most of our information today comes from the Internet, influencing what we know and our opinions, it is clear that we could be easily manipulated by technology to the advantage of those controlling the information we consume. Yet, if we look around us, technology continues to be developed without thinking about its risks. Without serious dialogue between different disciplines, and given that different stakeholders have conflicting interests, power structures will determine our future trajectories as they have currently.

Broken dialogues have even a deeper impact because different stakeholders try to advance their own interests. Sometimes these interests get embedded in ideas that are attractive and have become common knowledge while they are in fact often misleading. For example, consider the following statements: (1) Science uncovers the truth about the world—this gives scientists distinguished status and scientific knowledge becomes superior to other forms of knowledge—a position that has been at least contested.[23] Further, if science uncovers the truth about the world, it does not change it and this relieves scientists from thinking about the consequences of their discoveries. (2) Democracy means equal rights for all citizens—but most times decisions are made based on the interests of powerful minorities.[24] (3) Law serves justice—this gives lawmakers and those who control them supreme power. A law designed to serve a powerful minority becomes inherently unjust, creating unbelievable anomalies such as the differences in the punishment for different crimes.[25] (4) Innovation leads to happiness—so why is there so much stress in society today?[26] In addition, why is innovation used to addict people to products they do not need or even harm them?[27] Finally, (5) globalization benefits everybody

---

[22]Richtel (2014).

[23]Sarewitz (2016).

[24]Gilens and Page (2014).

[25]For example, California "three strikes" sentencing law as enacted in 1994, allowed punishing for food theft up to life sentence with documented people who got this punishment (e.g., http://www.sfgate.com/news/article/Stealing-one-slice-of-pizza-results-in-life-3150629.php or http://thegrio.com/2010/08/18/homeless-man-free-after-13-years-in-jail-for-stealing-food/); the law was amended in 2012 to make it less harsh, http://www.courts.ca.gov/20142.htm. In contrast, only one Wall Street executive was prosecuted and sent to 30 months in jail after the 2008 economic crises despite clear violations of the law. While this might not be a conspiracy to avoid procession, it may be a result of lack of knowledge or skills to deal with white-collar crimes that reflects priorities to invest resources in different aspects of the justice department, https://www.nytimes.com/2014/05/04/magazine/only-one-top-banker-jail-financial-crisis.html. Even the almost $190B fines and settlements paid by financial institutions following the crisis were shareholders money and not the executives themselves. The CEO of JP Morgan Chase even got a huge raise after closing the settlement of the bank. Other key bankers involved in the crisis received government positions in the Obama administration. All these reflect interests, power, but nothing resembling justice.

[26]Thielking (2017).

[27]A good example is the gambling industry that addicts gambler by design, Schüll (2014). Other examples include all "proper" designed websites or apps whose top goal is to keep people on their page as much as possible.

—but it turns out not to be this way for many workers in many countries.[28] We see how language creates powerful idioms, rather than truths, that serve the interests of those who advance them while precluding dialogue about their meaning. We see how controlling the language means power.

6.4 The PSI Matrix for Understanding Design

Broken dialogues lead to different interest groups maximizing their benefits, at best while ignoring others, but as indicated before, many times through exploiting others by design. Local maximizations by different groups will almost never lead to global optimum. This local–global (part-whole or self-community) distinction is a central concept in design. It permeates all aspects related to design including: (1) the spending on components vs. spending on their integration, (2) the needs of individuals vs. those of the community, (3) the personal skills vs. teamwork, or (4) focusing on a tool or a process vs. on organizational culture. Balancing this part-whole relation is key to designing. In order to design a better future, we need to understand better what is this designing skill that is at the heart of our problems and their solutions.

The PSI matrix is a framework that allows understanding designing and managing it toward success, where even what is considered as success is a product of the design process. The framework is borne out of understanding that even when we speak about designing; it is not a homogeneous endeavor. We need a trans-disciplinary framework to understand it. Clearly, any framework about design should deal with at least the following questions:

1. *What* is the problem or issue *we* are *addressing* by design?
2. *Who* is involved in *defining* the problem and *addressing* it?
3. *How* is the *problem* being *addressed*?

These seemingly simple questions embed some major complexities. For example, who is the "*we*" in the "*what*" question? How does this act bring together the people to form this "*we*"? How is the second question answered before knowing what is the "*problem*"? There are interdependencies between these questions and their answers, and it is unclear who has the responsibility, skills, methods and culture, and authority to address them. It is a situation where parts become wholes elsewhere and vice versa—it is a circular recursive situation.

The complexity goes even further as in order to answer the "*what*" question, knowledge from diverse disciplines is necessary including: engineering dealing with technology, sociology dealing with human needs or culture, and history dealing with the historical context embedding the problem. But there is hardly any dialogue between these disciplines.

---

[28]Goldsmith and Mander (1997).

To answer the "*who*" question, knowledge from at least psychology, sociology, and education is necessary to understand how people behave and work together even if they come from diverse backgrounds and cultures, what motivates them and how their capabilities and skills could be assessed and improved? To answer the "*how*" question, knowledge of management, economics, and even law must be exercised to create the working infrastructure that will enable all the activities required starting from problem identification, collection of participants, and infrastructure setup. Clearly, these disciplines hardly speak with each other and the recursive nature of design surfaces again.

Figure 6.1 describes the PSI matrix (see Footnote 15). The bottom level is the level that describes the daily activities we have concentrated on thus far, consisting of three spaces: problem (P) space that characterizes what is the problem; social (S) space that characterizes who addresses it; and institutional (I) space that characterizes how it is addressed. When dialogues break (for example, due to conflicting interests), or do not exist (due to lack of common language), the lower level malfunctions. This may lead to one group maximizing its benefits at the expense of others because not everybody is included in the "who," and the "how" creates an asymmetry of information flow. Malfunctioning of the lower level may also lead to wasting resources and creating solutions that are worse than the original problems.

In the case of such malfunctions, knowledge of design would tell us that a second level PSI is needed and constituted. Systems, companies, or even societies may have aspects of this level in the form of control systems, audit committees, or regulatory agencies, but again, we are forced to ask, what are the issues they address? who is involved, and how do they operate? We have at this level the same complexity we had at the first level PSI. The third and top level PSI needs to determine the vision, ethos, or "DNA" of the social system we are discussing, again with similar complexities of issues as the first two levels.

For a system, organization, or society to function well, all these levels, and the P's, S's, and I's in them have to be aligned and balanced while interacting with each other between and across levels. Metaphorically, this resembles a tensegrity structure (Fig. 6.2) where the rods are the PSI matrix elements and the strings connect them into a system and balance them. If one rod is extended beyond

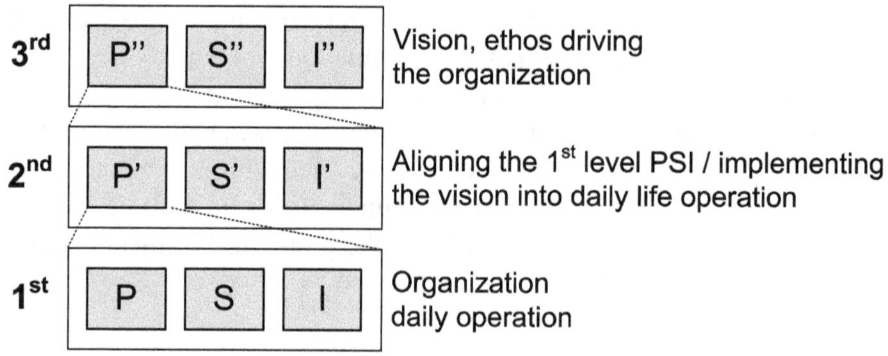

Fig. 6.1 The PSI matrix (see Footnote 15)

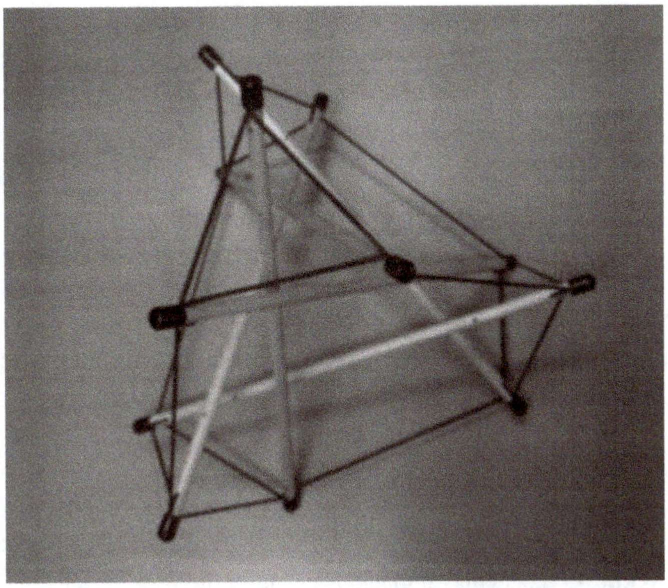

Fig. 6.2 A metaphor for the PSI matrix—a tensegrity structure (the structure was taken from the lab of the late Prof. Offer Shai)

proportion, representing a local maximization, the structure may fall apart. If all rods grow together, there is also a point where the strings break, suggesting that growth need to be accompanied by a structural change. Furthermore, some configurations with specific rod sizes may be singular. It is interesting to note that controlling the movements of such structure and checking its singular positions is complex[29]; consequently, projecting such issues onto complex systems, organizations, or societies may be overwhelming.

At this point, it is clear that even a parsimonious framework for understanding design, like the PSI matrix is very complex and even could be fragile. We can use the framework to understand how design situations could break or be used wrongly by some. We could use the framework to guide us from a particular state in these spaces to a desired state.

6.5 Epilog

Our reality is the result of design activities. We can only understand how we arrived at the present situation and move forward if we understand that the answer starts by improving the design capabilities of people or of society. We can only achieve this if we stir the course of education from disciplinary specialization at the least with some minimal inter or multidisciplinary collaboration, to truly create new languages and culture for transdisciplinary endeavors. For education to respond to all its levels this creates a serious challenge but there seems not to be any other way.

[29]Consider the following sources: Cahan and Shai (2015), Orki et al. (2012), Slavutin et al. (2018).

The design hypothesis and the PSI framework provide a first glance at what we need to study, further develop, and teach. It is outside the scope of this essay to go deeper into the framework's upper levels or attempt an explanation of why previous attempts at addressing the broken dialogue have failed. Analysis of different contexts such as organizational failures have been done (see Footnotes 14 and 15) and continue to be explored,[30] but in general, they are complex because they need a dialogue between numerous experts in diverse disciplines that are related to the PSI as were indicated before. The purpose of this essay was to raise the issue of broken dialogue again; to propose that design is at the heart of understanding past failures and present causes of difficulties; and to make the case that design offers the opportunity to design a better future. This requires that we educate people about it, become more aware of who tries to exploit it, understand how we can fail and be cautious and proactive about it, and engage in a collective dialogue about the role design plays in our daily individual and social lives.

Acknowledgements I thank my long-term friend and colleague Eswaran Subrahmanian for inviting me to the NSF-ESI Workshop at Carnegie Mellon University, for our almost daily discussions from which the material presented in this essay emerged, and for his comments on a draft of this essay.

References

Akao, Y. (1990). *Quality function deployment: Integrating customer requirements into product design*, (G. H Mazur, Trans.). Cambridge: Productivity Press.

Benkler, Y. (2006). *Wealth of networks: How social production transforms markets and freedom*. New Haven: Yale University Press.

Cahan, D., & Shai, O. (2015). Combinatorial method for checking stability in tensegrity structures. In *ASME Design Engineering Technical Conferences*, August 2–5, Boston, Massachusetts, USA, 2015.

Crawley, E., Malmqvist, J., Östlund, S., & Brodeur, D. (2007). *Rethinking engineering education, The CDIO Approach*. Berlin: Springer.

Diamandis, P. H., & Kotler, S. (2012). *Abundance: The future is better than you think*. New York, NY: Free Press.

Ferrara, E., Varol, O., Davis, C., Menczer, F., & Flammini, A. (2016). The rise of social bots. *Communications of the ACM, 59*(7), 96–104.

Fuller, B. (1960). *World game*. https://www.bfi.org/about-fuller/big-ideas/world-game.

Gilens, M., & Page, B. I. (2014). Testing theories of American politics: Elites, interest groups, and average citizens. *Perspectives on Politics, 12*(3), 564–581.

Goldsmith, E., & Mander, J. (1997). *The case against the global economy and for a turn towards the local*. California: Sierra Club Books.

Kramer, A. D., Guillory, J. E., & Hancock, J. T. (2014). Experimental evidence of massive-scale emotional contagion through social networks. *Proceedings of the National Academy of Sciences, 111*(24), 8788–8790.

---

[30]One present study uses the PSI matrix to understand why different projects of creating sustainable communities in Israel succeeded or failed.

Orki, O., Ayali, A., Shai, O., & Ben-Hanan, U. (2012). Modeling of caterpillar crawl using novel tensegrity structures. *Bioinspiration and Biomimetics, 7*(4), L 046006.

Papanek, V. (1971). *Design for the real world: Human ecology and social change*. New York, NY: Pantheon Books.

Perrow, C. (1984). *Normal accidents: Living with high risk technologies*. New York, NY: Basic Books.

Reeves, R. V. (2017). *Dream hoarders: How the American upper middle class is leaving everyone else in the dust, why that is a problem, and what to do about it*. Massachusetts: Brookings Institution Press.

Reich, Y., & Shai, O. (2012). The Interdisciplinary Engineering Knowledge Genome. *Research in Engineering Design, 23*(3), 251–264.

Reich, Y., & Subrahmanian, E. (2015). Designing PSI: An introduction to the PSI framework. In *Proceedings International Conference on Engineering Design, ICED'15*, Milan, Italy, 2015.

Reich, Y., & Subrahmanian, E. (2017). *The PSI Matrix—A Framework and a Theory of Design*, ICED'17, Vancouver, Canada.

Richtel, M. (2014). *A deadly wandering: A tale of tragedy and redemption in the age of attention*. New York, NY: William Morrow.

Sapolsky, R. M. (2017). *Behave: The biology of humans at our best and worst*. London: Penguin Books.

Sarewitz, D. (2016). *Saving science—The New Atlantis* (pp. 4–40). Berlin: Spring.

Schüll, N. D. (2014). *Addiction by design: Machine gambling in Las Vegas*. Princeton, NJ: Princeton University Press.

Slavutin, M., Sheffer, A., Shai, O., & Reich, Y. (2018). A Complete Geometric Singular Characterization of the 6/6 Stewart Platform. *Journal of Mechanisms and Robotics*.

Tadmor, Z. (2006). Redefining engineering disciplines for the twenty-first century. *The Bridge, 36* (2), 33–37.

Thielking, M. (Feb 6, 2017). *A dangerous wait: Colleges can't meet soaring student needs for mental health care*, STAT. https://www.statnews.com/2017/02/06/mental-health-college-students/.

Tierney, T., & Baumeister, R. (2011). *Willpower: Rediscovering the greatest human strength*. London: Penguin Books.

Winner, L. (1980). Do artifacts have politics? *Daedalus, 109*(1), 121–136.

Wolfram, S. (2002). *A new kind of science* (Vol. 5). Champaign, IL: Wolfram Media.

Wu, T. (2013). The case for less: Is abundance really the solution to our problems? *New Republic*, April 23. 2013. Retrieved April 1, 2017, from https://newrepublic.com/article/112858/abundance-really-solution-our-problems.

Zuboff, S. (2015). Big other: Surveillance capitalism and the prospects of an information civilization. *Journal of Information Technology, 30*(1), 75–89.

Chapter 7
Engineering Design and Society

Shyam Sunder

As a discipline, engineering has remained rooted in mathematics and natural sciences throughout recorded history; as a practice, it is social at its core. A large part of engineering is, and always has been, a craft based on imagination and ingenuity combined with day-to-day incremental learning from practical, shared, inherited, and emulated social experience. Motivation for most engineering is to improve our individual and collective lives, but its consequences can extend far beyond the intent.

For example, fences surrounding early human dwellings protected them from marauding animals and people; these very fences shaped our civilization and society through many consequences not anticipated at the outset. In exploring the linkages between engineering and society, it is useful to keep in mind both intended uses and unintended consequences that arise or are discovered later through complex social interactions. Only a part of engineering's ultimate impact on society can be deliberately targeted, or even imagined, at the outset.

Objects of natural science and products of engineering are not sentient. Absent self-consciousness, such objects are not aware of any laws or regularities that we may discover about them, or of human purpose behind the artifacts or processes that may be designed by engineers. This ignorance lends robustness to knowledge in the natural sciences and the properties of engineered products.

Social science is a more recent development than the society that it seeks to document and understand—perhaps as recent as Aristotle (4th century BCE), Kautilya (4th century BCE), or Ibn Khaldoun (14th century AD). As a discipline,

An earlier draft of this paper was prepared for and presented at the National Science Foundation Workshop on Engineering a Better Future: Interplay between Engineering, Social Sciences and Innovation, Carnegie Mellon University, April 15–16, 2016.

S. Sunder (✉)
Yale School of Management, Yale University, New Haven, CT, USA
e-mail: shyam.sunder@yale.edu

E. Subrahmanian et al. (eds.), *Engineering a Better Future*,
https://doi.org/10.1007/978-3-319-91134-2_7

social science faces a more difficult challenge. Since societies comprise self-conscious human beings endowed with at least a degree of free will, the explanatory power of any laws or regularities in social science tends to be lower. Moreover, self-conscious humans are prone to alter their behavior in response to any regularity social science may discover and reduce the robustness of such laws to their own discovery.

Bringing social science into engineering design has its pros and cons. Identifying diverse interests in engineering design is only the first step. Legitimate interests must be separated from interests that are best ignored and set aside. The remaining interests have to be assigned some reasonable weight in a space of incommensurate interests. Will engineering design still be possible when all legitimate interests are given reasonable weight? Will the results satisfy everyone, or be acceptable to all? New York did not extend its subway system to JFK airport, but built a separate line, presumably to accommodate the interests of taxi and limousine companies and drivers. Is that a better solution for society? Might it have been better instead to design the best "engineering" solution for transportation to JFK airport and deal with the political fall out afterward?

Differences in, and mutual engagements of, engineering and social science are interesting and important. Engineers design ways and artifacts to do things better; social scientists concern themselves with understanding how things are, and why. Engineering measures itself by its ability to change lives; intervention is less welcome in social science. Unlike social science, engineering is concerned with objects. Absent consciousness in its objects, the principles of engineering design are robust to variations in time and space; social science has more difficulty identifying regularities that are robust even to their own discovery. The successes of engineering are easily seen in society's high regard for them, but social engineering evokes skepticism and suspicion.

These disciplinary differences are rooted in the strategic use of information by conscious (not material) objects of study. However, in practice, engineering and society interact intensively and shape each other. Through its myriad innovations—fire, weapons, tools, and processing—engineering has radically redefined the very nature of human society since prehistory and it continues to do so. In this way, modern society is an offspring of engineering, while much of engineering is also an outcome of social organization, human relations, and interactions. Engineering is a deliberate, purposive activity of individuals or groups seeking their objectives. Social phenomenon, on the other hand, is conceptualized at an aggregated level as a consequence of individual behavior. Engineering is a practice; society is the result of many such practices.

This paper explores the mutual relationship between engineering and society and the ways in which social science influences the effectiveness of engineering design and interventions. This examination of design and social science is organized into six sections centered on information.

1. **Information and Design**: *Science creates information; engineering encapsulates the information discovered or extracted by science into the design of artifacts to suit our wants, and thus it redefines work, lives, and ultimately society.*

Scientific knowledge (natural as well as social) expands our information set about the world. When we use this information to effect change—often in an attempt to improve our work and life—we practice engineering. Society (an interactive network of individuals) and engineering are in this sense inseparable. It is useful to explore the nature of this linkage by probing into the nature of work.

Work involves an entwined matrix of physical action and information processing. The consequences of rearranging action with the use of knowledge pervade society. The act of writing, for example, consists of obtaining a piece of paper, finding a writing instrument, and moving one's hand (all actions) guided by information about the location of the writing materials, what words or figures to draw, and how to draw them. Technology can change these relationships; consider the example of the shift from handwriting to typewriting in the 19th century. The introduction of the typewriter replaced the need for knowledge about how to form letters and words using a pen or pencil with the knowledge about how to make and operate a typewriter. Typing technology retains paper, but replaces a pencil with a ribbon and the knowledge of letter formation with preformed typefaces. This technological invention comprises a much larger body of knowledge including conceptualization, design, manufacture, and distribution.

Learning to write in grade school can now be confined to character recognition and exclude character formation.[1] Legible writing is speeded up and is read faster to save time at both ends. Writing is standard in form and character spacing. New classes for learning typewriting in schools and businesses appear and the "typist" appears as a new business professional. The potential for errors creates a demand for specialized proofreaders and symbols for copyediting. New industries for making and repairing typewriters and keeping them supplied with ink, ribbons, and replacement parts are born. This generates a demand for labor, materials, manufacturing machines, space, financial capital, power, and entrepreneurs. Supply lines of production are adjusted to fit the new technology. Today, typewriters have been replaced by computers and printers whose origins and consequences have little continuity with the technology of "writing" they replaced. Engineering innovations ripple through all aspects of society.

Many other examples of how engineering innovations shape our work and all aspects of society exist: the replacement of horse carts by trains and automobiles radically altered not only transportation but also the design of cities and towns; it led to the creation of roads and rails, new sources of energy, businesses, social interactions, manufacturing, retailing, skills, professions, colleges, educational

[1]The average quality of handwriting is said to have deteriorated during the recent decades, just as the introduction of calculators is said to have trimmed the ability to calculate. Whether elementary schools should allow/encourage mechanical aids for writing and calculation remains controversial.

programs and diplomas, and academic disciplines. Not every engineering consequence is positive however, new technologies also give rise to new methods of waging war, killing, exploitation, and poverty.

Design gives body to information about the laws of nature in the material world and to our wants in the social world. The ancient human desire to fly like a bird only ended in broken bones or tragedy, until the Wright Brothers brought their perspective of the laws of nature to build their flying machine. A century later, every component of a jet aircraft encapsulates vast amounts of information to realize that dream for the masses.

Civilizations accumulate information and the consequent capacity to design and build both physical and institutional artifacts. Revolutions transform or destroy the accumulated artifacts in order to begin anew with a clean slate.

Artifacts alter existing designs (well beyond their makers' intent) in stages or layers. When a design is sufficient and successfully executed, for example, a bicycle, the first layer defines the intended change—easier and faster travel on the new machine. Increased demand for use adds subsequent layers—the straightening and paving of cow paths, the design of houses, the layout and density of towns, training people to build and maintain the bicycles, places to park, store, and secure the new machine, demand for bicycle parts, and movement of materials and people to support the technology and economics of the new artifact. Every layer calls for new information, possibly discarding some information, skills, facilities, parts, and organization previously needed for the displaced artifacts (e.g., saddles, food, shelter, and care for horses).

The consequences of engineering pervade all aspects of our work, redefining information and skills needed for a given function thus generating new kinds of work, while making others obsolete. It also redefines how we live, spend our time and material resources, relate to one another, and what is produced, where, and by whom. It redefines who consumes it and consequently trade and commerce, the distribution of wealth, and the social and political architecture that supports us. However, most if not all of these social consequences of technology are unknown and possibly unknowable to the inventor of the artifact at the time of its design and introduction. Henry Bessemer could not have predicted the global consequences of the large-scale production of iron from the blast furnace he designed to smelt iron ore; his purpose was local, but the consequences of his design were far-reaching in time and space. In economics, the distinction between partial and general equilibrium recognizes this layering of consequences.

2. **Replication and Scale**: *The replication of material is expensive in effort and time; the replication of information is free.*

Transforming lumber into ten chairs of a given design takes a carpenter almost ten times as long as it takes to make one chair; however, the design of the chair (and the information encapsulated in its design) can be replicated endlessly without significant additional effort. In other words, the cost of production in transforming materials is variable, but the cost of information embedded in the design is mostly fixed (with respect to the number of units to which it is applied). By incorporating more information into product design, the (hardware or software) engineer substitutes variable cost by fixed cost. Increasing the fixed costs does not make sense unless we consider the economies of scale. No matter how large the additional fixed cost, increasing the number of units always reduces the cost per unit. When the number of units demanded and produced is raised sufficiently, the unit cost falls below the original variable cost.

Falling unit costs arising from engineering design, which appear to be a blessing at first, have extensive economic and social consequences. Not all of these consequences increase social welfare. The least-cost mode of production is a monopoly, but the monopolist may exploit market power to sell fewer units of the product at a higher price to maximize profit (in comparison to competitive equilibrium levels of price and quantity). An economy dominated by a monopoly also has less product diversity and fewer opportunities to learn and innovate.

Unless they are carefully controlled, a monopoly's consequences in the form of stifled innovation, high prices, and product rationing are norms, not exceptions, and arise from embedding information to improve the design. Indeed, most major technology firms in recent decades, including Westinghouse, General Electric, IBM, AT&T, Microsoft, and Google (renamed Alphabet), have been restrained by antitrust authorities in the US and the EU. Database technology introduced massive economies of scale in accounting and other operating costs of financial services in banks, insurance, and mutual funds. The cost of developing software to maintain a thousand customer accounts is not much lower than the cost of software for hundreds of thousands of accounts. These economies enabled the consolidation of financial service firms and the growth of behemoths such as J.P. Morgan Chase which are not only too big to be allowed to fail but also too big to jail in democratic polities where donations finance political election campaigns. It has been argued that the Atlantic Financial Crisis of 2007–11 may have been rooted, at least in part, in information economies of software design to operate financial service firms (Chaudhri et al. 2018).

3. **Choice Criteria**: *Engineering, economic, social, and moral interpretations of efficiency form a hierarchy, and their precision and completeness decline with the increase in relevance to society.*

Efficiency, over the spectrum of its varied forms, is the criterion for making decisions and evaluating outcomes. In engineering, the efficiency of an artifact is defined as the ratio of its output to input, each measured by a single variable, e.g., a vehicle's miles per gallon of gasoline or the percent of the metal content extracted from ore.

In the presence of multiple inputs and/or outputs, it is useful to convert each of them into a common unit before calculating the efficiency of the artifact as the output/input ratio. For example, if the inputs consist of multiple types of fuels, they might be converted into equivalent British Thermal Units (BTUs) to calculate efficiency. When such equivalence across incommensurable inputs or outputs is not meaningful, it is useful to switch to economic efficiency by converting various inputs and/or outputs into units of money before calculating efficiency. The cost in dollars per gallon of gasoline dispensed at the gas station is calculated from the cost of many inputs for producing, transporting, storing, and delivering the fuel to car tanks. Some economic measures of efficiency leave out variables which cannot be converted into money (human life, morality, beauty, etc.).

Both engineering and economic measures of efficiency are computed from a single point of view—the party who contributes the inputs and receives the outputs of the process. In social settings, two or more points of view are present, and each party views the process on the basis of efficiency of resource costs and benefits accruing to that party. The same process can be attractive for some, but harmful to others. A Pareto efficient process is one that benefits at least one party without hurting any of the others. Unlike engineering and economic concepts of efficiency, the Pareto criterion is incomplete because generally it cannot rank all of the processes in order of their attractiveness; it is possible for some processes to be neither more nor less Pareto efficient when they confer advantages and disadvantages on different parties.

Moral criteria lie on the spectrum of efficiency concepts beyond engineering, economic, and Pareto considerations. The interests of some parties may be excluded on the grounds of being morally inadmissible. In a society with property rights, better locks may be morally efficient even though they may deprive thieves of their livelihood. The identity of parties' whose interests are admitted to efficiency calculation is a moral, not technical, judgment. Moral criteria may also include the cross-sectional distribution of the consequences of an engineering choice borne by various members of society. For example, fairness may lean in favor of equality or toward the weaker sections of society.

This hierarchy of engineering, economic, Pareto, and moral efficiency takes us from uniqueness, clarity, hardness, and completeness at one end toward multiplicity, ambiguity, judgment, and incompleteness at the other. Arguably, it also takes us in the direction of greater relevance to society and creates a difficult-to-resolve tension for engineering design. Which criterion or criteria can one use to identify which engineering designs are better for society? Can we determine the chosen efficiency of the proposed designs ex ante, especially in the presence of risk and uncertainty, and the differential spread of costs and benefits over time? If so, how?

4. **People**: *As self-aware, thinking, and sentient beings, we respond to all-natural and engineered changes in our environment; our faculties to learn, conjecture, and anticipate lead to the indeterminacy of, and difficulty in predicting outcomes in social systems.*

In predicting the social outcomes of engineering interventions, we face major hurdles. First, unlike inanimate objects, human beings remember, conjecture, anticipate, learn, strategize, and react to changes in their environment. Each act of engineering alters the environment and induces changes in the behavior of people in that environment. These changes can be direct (e.g., one's commuting route after construction of a new bridge) or indirect (e.g., living farther from the place of work after construction of a faster highway or commuter railway line). Engineering-induced indirect changes permeate society. Yet, it is difficult for the engineering designer to anticipate even a significant proportion of these consequences of design in a complex system.

Second, humans deliberately create new ideas and things that did not exist or had not been imagined before. In choosing a courses of action, we exercise free will, making it difficult to predict what an individual will do on the basis of past observations. Choice theory seeks to explain human behavior by postulating unobservable preferences, which in turn are estimated from observed past behavior. Absent a modicum of stability and generalizability in time and space (which would allow one to postulate that the data on behavior at one time or in one situation can be used to estimate preferences applicable to behavior in another time or place), choice theory risks becoming circular, and therefore moot in furnishing useful explanations or predictions of individual behavior.

5. **Dispersed Information**: *The outcomes of extended order cannot be predicted with the limited access to the microlevel knowledge in possession of individuals (Hayek: The Fatal Conceit).*

A third hurdle facing engineers in trying to predict the social consequences of design intervention arises from the dispersed nature of information in society. A large part of information comes to individuals as a by-product of living daily lives observing and doing things. Gathering all this information at a central location to make it available to a decision-maker is difficult, if not impossible. Asking individuals to share such information raises the possibility of strategic manipulation, withholding, and refusal to reveal the truth. Hayek (1991) points out this fundamental and pervasive handicap of central planners. It also applies to engineers trying to design an artifact for a targeted social consequence. Determining where and what kind of highway or bridge is built, and how much tax money is spent on elementary education or public health, are examples of this difficulty. Nelson (1977) points to the shortcomings of the economic, organizational, and scientific approaches in helping make normative judgments on public policy. The metaphor in the title of his book refers to the relative ease of putting man on the moon as compared to the difficulty of improving the education of children raised in poverty.

6. **Evolution**: *Engineering designs are mutations; some disappear in the emergent process of social evolution; others are adapted and lose their identity; some survive in a secure niche to furnish ambient variation, while a few spread widely to redefine society.*

From the point of view of social consequences, engineering design can be seen as a biological mutation in genes in our macro domain. To the best of our knowledge, biological mutations are random errors due to copying or exposure to radiation, and follow no known pattern. In design, the creative spark and imagination is the "random" stimulus. Most changes in design may have no detectable consequences. Some mutations diminish the fitness of the organism in its environment and tend not to survive over time. Others may survive but are adapted over time through subsequent mutations to the point where they lose their distinct identity. Many surviving mutations, like Darwin's finches on the Galapagos, find and survive in their own narrow niches. Radical innovation does not occur in biology—the gills of a fish are not replaced by lungs in its offspring in a single mutation.

Hyperbole aside, almost all engineering designs are evolutionary, not revolutionary. Like DNA, the design also encodes information from past experience and newer insights and hands it to the next generation. This chain of encoding an increasing amount of information in surviving design is a function of social, political, and economic organization and stability. The lost engineering of fallen civilizations and the puzzlement during the dark ages that follow are the norms, not exceptions; the metallurgy of the massive iron pillar of Delhi, which remains rust-free after 15 centuries, had been lost to mankind until the early twentieth century.

Concluding Remarks

The distance between various disciplines and their corresponding practices varies. This distance is narrower in engineering, but relatively large between social science on the one hand and society on the other hand. The practice and discipline of engineering interact more intensively and contribute more to each other. Obviously, society shapes social science, but the influence of social science on society is weaker to the extent that, in some quarters of social science, interest in society itself starts and ends with being the object of their scholarship and source of observations. The practice of engineering is purposive while society is an aggregate-level phenomenon to which purpose is not easily ascribed. Engineering design often serves as an intervention and alters social relationships that social scientists study. The adoption of an engineered product becomes part of the social fabric and an object of study in social science. Known properties of this changing social reality are not as robust as the objects of engineering.

Items (4) and (5) in the list above illustrate the difficulty of predicting the social consequences of engineering designs or designing an intervention to achieve a given social consequence. Their social level consequences, especially in the second and higher layers, become known only with experience after the design is implemented. These social consequences emerge from myriad interactions of sentient individuals with the design, and with one another. The complexity of these individual interactions can be analyzed statistically or simulated if we have access to sufficient knowledge of rules of their dynamics; however, the latter is rarely possible.

Whether or not social science can serve to inform better engineering design depends on our ability to understand and predict the social consequences of design. The social effects of engineering design may vary by context and a design targeted for one context may not have its intended effect in another. Beyond traditional engineering, products such as robots with limited self-consciousness and social ability may change the meaning of social relationships. When products have the potential to be autonomous agents, social science will face new challenges.

Fortunately, all is not lost. Gode and Sunder (1993) report evidence that it is possible for markets to have certain well-defined and predictable properties even if the behavior of individual agents is unpredictable. Identifying corresponding propositions about interaction between the design and society may be a fruitful line of inquiry. Until we learn more, we have to be modest about the ability of social scientists to help design better physical or institutional artifacts on the basis of *their science*—the documented laws of their discipline.

Social scientists can conduct surveys before the designs are implemented or study the reactions and consequences afterward by gathering and analyzing data. But it would be unwise to expect much predictive power from social science about the effect of these designs on society. With rare exceptions—e.g., Ibn Khaldoun (1958 [1377])—the record so far does not support much confidence, especially in establishing causality and out-of-sample predictive power. Engineering has not done too badly being on its own. The grass in the social science yard may seem greener than it is from the engineering side of the fence.

Consequently, one may argue that social science is robust only to a certain extent as it tends to study how things are when technology and social context change. So it may be that we are chasing a chimera of science in social science rather than seeing its role as one of studying how things are in a socio-temporal context while looking for "some invariants" that could help us predict the change that an engineered product may have. In some ways, it is like regulating a new technology; regulation lags the introduction of technology because it is difficult to predict any consequent social failures, especially emergent unanticipated interactions. However, social science may help us know the types of social failures that we may have encountered before in the introduction of prior technologies. For example, one could argue that the Wright Brothers' airplane is not the same as a Boeing jet in spite of them sharing the physics of aerodynamics. They do not share the physics of a jet engine which was developed in a different society and technological environment many decades later.

We can ask people potentially affected by design to contribute suggestions and an alternative design. Field or beta tests and focus group assessments of new products and services are a common practice. This can yield unanticipated glitches and new ideas for the designer. It also renders radical innovation less likely to succeed if it does, or appears to, hurt the interests of some well-organized interest group. The nonexpert users can add a valuable perspective, but they also lack the big perspective that a Steve Jobs for example may bring to redefining the cell phone instrument. Holbrook (2016) suggests that new airports around the world seem to be made for everyone—cities, airlines, retailers, architects—but passengers. Given the opportunity to alter a design, will people at large help improve engineering? We do not yet know the answer.

Alternatively, we can take a narrower engineering efficiency perspective at the design and construction stage and let the acceptance of design to society be sorted out after the fact in a manner parallel to how nature accepts and rejects biological mutations. Of course, the rate of rejection of biological mutations is high and may be lower in engineering design. Experience with a given design will yield feedback to the designers, who may include it in the next generation of designs; however, this manner of using feedback is slow and can be costly in the long run.

Acknowledgements Without implicating them, I am grateful to E. Subrahmanian, participants in the Workshop, Manjula Shyam, and Elizabeth Viloudaki for their helpful comments and suggestions.

References

Chaudhri, V., Magnusson, G., & Sunder, S. (2018). *Economies of scale in banking*. Yale School of Management Working Paper.

Gode, D. K., & Sunder, S. (1993). Allocative efficiency of markets with zero intelligence traders: Market as a partial substitute for individual rationality. *The Journal of Political Economy, 101*(1), 119–137.

Hayek, F. A. (1991). *The fatal conceit.* Chicago: University of Chicago Press.

Holbrook, C. (2016, April 6). Airports, designed for everyone but the passenger. *The New York Times*.

Ibn Khaldoun. (1958) [1377]. *The Muqaddimah: An introduction to history* (3 Vols., Translated to English by F. Rosenthal). New York: Pantheon. http://www.muslimphilosophy.com/ik/Muqaddimah/.

Nelson, R. R. (1977). *The moon and the ghetto: An essay on public policy*. New York: W. W. Norton.

Chapter 8
The Cult of Innovation: Its Myths and Rituals

Langdon Winner

Philosophers and scholars of rhetoric point to the significance of what they call "god terms," concepts that have a certain "inherent potency." God terms sweep up whole periods of history as nations and cultures strive to reach a higher state of being Richard (1953). From the late 1700s and well into the early decades of the twentieth century, for example, a prominent focus of inspiration was "Revolution," an idea that heralded sweeping social, economic, and political upheaval with the expectation of wonderful outcomes. By the same token, a fascination with "Progress" during the Enlightenment promised inevitable improvement in human knowledge and its beneficial, universal applications. Of course, a perennial favorite in the United States has long been that of "Frontier". While the specific location and meaning of the zone have changed over the years—from the geographical expanse of a great continent, to the horizons of modern science, to the space program of the 1960 geographical, to the realm of cyberspace widely celebrated in recent decades—Americans have always looked longingly toward the next "Frontier" looming over the horizon.

A popular god term today has become an object of worship in universities, think tanks, corporations, Wall Street brokerage houses, and in the dreams of our social elites. The concept is widely associated with originality, vision, inventiveness, success, wealth, fame, personal virtue, national prosperity, and cultural vitality, and features are widely understood to express the aspirations and accomplishments of twenty-first-century societies at their very best. For many people, concept has become the source of their deepest spiritual aspirations and yearnings for transcendence. In fact, it is not an exaggeration to say that today's central "god term" has begun to resemble a cult with ecstatic expectations, unquestioning loyalty, rites of veneration, and widely echoing exhortations of groupthink.

L. Winner (✉)
Department of Science and Technology Studies, Rensselaer Polytechnic Institute, Troy, NY, USA
e-mail: Langdon@rpi.edu

E. Subrahmanian et al. (eds.), *Engineering a Better Future*,
https://doi.org/10.1007/978-3-319-91134-2_8

The god term I have in mind is, of course, "innovation." The word derives from the Latin *innovare*, which means "to renew" or "to restore." In everyday speech, the word has come to mean something like: the activity of bringing new things into being that will generate sweeping renewal throughout the world. Thus, "innovation" is an inherently, overwhelmingly positive concept, a quality that may explain why so many of our contemporaries are transfixed by its very mention. In fact, it seems that a person would have to be nearly crazy to call the phenomenon into question at all. How can one criticize something that is *inherently good*?

To explain the astonishing popularity of The Cult of Innovation one need look no further than the aura of personal creativity that surrounds it. People imagine themselves working with new ideas, starting new enterprises, revitalizing regional economies, playing a leading part in creating what they imagine to be a lively, engaging social future. Today it seems that every person, every organization wants to be associated with "innovation."

In many ways, the study and promotion of innovation have become a dynamic growth industry. There are now research centers, university departments, academic journals, book series, and the like engaged in an intensive quest to find the right recipe, the right mix of ingredients, exactly the right alchemical formula, to make it all happen. In deliberations about business strategy, public policy, and personal career planning, questions of this kind have become primary concerns. Which kinds of corporate culture, which varieties of university education, which programs of government support, and which social and psychological traits are the ones best suited to fostering innovation? Is advanced research and development the key? What varieties of management are most helpful? Isn't new technology the true fountainhead? Or does the magic arise from interdisciplinary or cross-cultural teams? What about the role of the arts and humanities in the alchemical mix? And what role should government play in the search for the new philosopher's stone?

Recommendations for promising recipes and methods often begin in Silicon Valley, Seattle, or Cambridge and move on to and recommend ingredients and likely to work in one's own corner of the world. We often hear that a particular firm, industry or whole nation has flourished because it found a distinctive path to the promised land. In contrast, we are sternly advised that conventional practices or policies—e.g., taxation or attempts at government steering and planning—are bound to be destructive because they tend to put a brake upon innovation, a most despicable thing to do.

I am certainly not immune to this way of thinking and talking. I now teach in a university undergraduate program Design, Innovation and Society, an outgrowth of an earlier cross-disciplinary program in Product Design and Innovation. We have our own favorite methods, recipes, and ways of motivating bright young women and men. Indeed, many of my own questions here stem from a simple desire to understand: What is this wonderful "innovation" process we all keep chattering about?

8.1 Gadgetology

At the level of popular culture, one feature of the Cult of Innovation I find fascinating is its veneration of the gadget. For many people in young people in colleges and universities, in Silicon Valley firms, and the like, the effective goal appears to be the search for the next iPod, iPad, or ingenious "app" for one's smartphone. Among those, I regularly talk with about such matters, if one asks—"What's the goal of the innovation you're working on?"—a common answer seems to be, "Well, it's a hand held digital device, a lot like an iPhone only different." "Or it's a smartphone app that helps people who want an easier way to…".

A while back I did an Internet search for lists of the most notable innovations of 2015.[1] Here are some of hundreds of notable examples from newspapers, magazines, and websites:

The HTC Vive: – a virtual reality headset with a pair of 1080 pixel viewing screens;

The Microsoft HoloLens – a camera equipped headset with augmented virtual reality capabilities for the gaming and educational markets;

Sony Smart Btrainer – earpieces that play music, monitor one's heartbeat and track data from one's daily exercise;

BB-8 A Star – a rolling Star Wars droid you can own;

6Sensor Lab Nima – a portable gluten sniffer (key breakthrough in America's war on gluten).

Not all of the "innovative" products were electronic in character. For example, the Nike Zoom Soldier Flyease 8 is an athletic shoe one can tie with one hand. What the other hand might be doing is not specified in the company's advertisements.

At the top of *Time Magazine's* list of the most notable inventions of the year was the Hoverboard Scooter. Shortly after it received this award, alas, the reputation of the Hoverboard collapsed when risks of fire and explosion caused by battery failure, forcing the recall of a reported half million of these nifty devices.

It is not always clever stand-alone gadgets that receive the applause. Some of the more fascinating innovations emerge as whole genres or networks, elaborate projects with a great many applications. "The Internet of Things" is an ongoing promotion of this kind. Even that, I am advised, may not be sensational enough. A friend who works in a firm devoted to innovative design tells me that the latest wrinkle is "The Internet of Caring Things." In a society that seems to care about us less and less, we'll have networked things that care more and more.

A notable example is Cuptime, a plastic cup that "connects wirelessly with a cell phone, allowing consumers to monitor their water intake." Obviously, this will be a

[1]Among the websites I visited were these: "The 25 Best Inventions of 2015," TIME Staff. http://time.com/4115398/best-inventions-2015/. "The 100 Greatest Innovations of 2015," Popular Science Staff, October 23, 2015. http://www.popsci.com/100-greatest-innovations-2015.

godsend that will greatly improve the quality of everyday life.[2] Of course, one could also just watch how much they are drinking. But that wouldn't be especially "innovative," would it?

Yes, it is impressive to see such ingenuity and devotion lavished upon the marvelous products and smartphone applications flooding the market. Yet one has to wonder, do we find ourselves living on the Planet of the Apps?

8.2 Elitism

Closely associated with Gadgetology is a penchant for elitism—inquiries whose likely outcomes are geared to the world's wealthy and already best served: well-to-do consumers and global firms. Because this is a signature feature of prevailing practice and because some of those who aspire to become "innovative" also have more noble goals in mind, there are now scattered efforts to disentangle high tech creativity from its emphasis upon the needs of the wealthy few. Implicit misgivings about elitism can be seen in creation of centers and programs that find it necessary to modify the names for what they are doing. The Center for Social Innovation at Stanford University wants us to understand that it "envisions a networked community of leaders… to build a more just, sustainable and prosperous world."[3] There are now several centers for "social innovation" around the U.S.A., ones that shift the focus of their inquiries beyond mere money and marketing. Other initiatives—indeed some favored by my own colleagues—are eager to point out that what they are doing actually is "good" or "ethical" innovation—technical and/or policy change aimed at the world's poor people or some other admirable causes. As laudable as such projects may be, they also tip the hand of "innovation" manias in common practice.

8.3 Benign Innovation

Despite these misgivings, it's clear that there are many of the kinds of creativity called "innovation" that truly are well worth celebrating. One could mention contributions that change, enliven, enrich, and extend long-standing and practices in the crafts, medicine, technology, music, and countless other fields of human endeavor.

---

[2]Internet of Caring Things, April 2014, Trend Briefing. http://trendwatching.com/trends/internet-of-caring-things/.

[3]Center for Social Innovation, Stanford Graduate School of Business. https://www.gsb.stanford.edu/faculty-research/centers-initiatives/csi.

At the top of my own list would be the astonishing career of Miles Davis, a great innovator in twentieth-century jazz. Beginning in the 1940s as a rather ordinary player in the style of bebop, Miles soon moved on to a more smooth, mellow harmonically complex style known as cool jazz. During the decades that followed, he again and again changed his approach to music, initiating or contributing to hard bop, orchestral jazz, jazz/rock fusion, hip-hop, and other styles.

Miles never rested on his laurels. At the pinnacle of success of one pioneering style, he abruptly would drop it and move on to something else. Thus, he left to the history of jazz a long sequence of stunning, successful, highly generative innovations. Yes, he intervened to interrupt and redirect the ways the music was conceived and played. But he also connected and uplifted, steps in the process often overlooked or even strongly disparaged these days.

Another creative soul of this kind, working in a much different domain, is Alice Waters, restaurateur and advocate of new ways of growing, cooking, and eating foods in the U.S. and around the world. Beginning in 1970s with her restaurant Chez Panisse in Berkeley, Waters took aim at the prevailing culinary practices of the time—overcooking, too much gravy, too much grease, etc.—and introduced methods that emphasize fresh, locally grown, organic ingredients, carefully prepared in a direct, tasty manner, a way of cooking that came to be known as the California Cuisine. Eventually, her "innovations" became a model for transformations in restaurant and home cooking that swept the country (and much of the world) during the decades that followed.

To my way of thinking activities and projects that modify and renew traditions and instruments of practice, might be called graceful or benign innovation. What characterizes them, in my view, is that they usually, deeply respect what came before and yet chart new, challenging, fruitful possibilities. The old traditions are not trashed, but modified, gracefully unfolding into something new.

8.4 Smash and Grab

Of course, in today's understandings of technological, financial, and corporate strategy, it is not the model of benign innovation that is the object of fascination and advocacy. Instead, the models praised are ones that involve deliberate social, cultural, and material violence—often described as "destruction" or "disruption." Far from respecting and building upon a tradition of tools and methods, the recommendation is to plunge ahead, discrediting, smashing up rapidly replacing what came before, usually with a narrow set of motives in mind—corporate profit and market capture.

While the term innovation has long had a long history in European languages, it is possible to date its emergence as a key concept in thinking about technology policy to the middle twentieth century, especially in the writings of the Austrian thinker Joseph Schumpeter, who eventually taught economics at Harvard. Pondering the dynamics of modern capitalism and in particular the ways in which new industries replace old ones and new product's replace functionally similar products of earlier times, Schumpeter proposed the notion of "creative destruction." He wrote, "The fundamental impulse that sets and keeps the capitalist engine in motion comes from the new consumers' goods, the new methods of production or transportation, the new markets, the new forms of industrial organization that capitalist enterprise creates." Schumpeter emphasized the dynamics of a "process of industrial mutation ... —that incessantly revolutionizes the economic structure from within, incessantly destroying the old one, incessantly creating a new one. This process of Creative Destruction is the essential fact about capitalism" (Joseph 1950).

Although Schumpeter's term seemed new and catchy at time, an entirely similar idea had been around for a long while. In 1848, Karl Marx and Friedrich Engels observed that "The bourgeoisie cannot exist without constantly revolutionizing the instruments of production. ... All that is solid melts into air, all that is holy is profaned."[4]

In recent years, the idea that change involves deliberate attack redefined under the rubric, "Disruptive Innovation," a theory and strategy widely promoted in business schools, Silicon Valley, and on Wall Street. Prominent spokesman for this tactic is Clayton Christensen of the Harvard Business School. Christensen's method is to locate existing sources of value contained within existing fields of endeavor—communications, transportation, health care, hotels, education, etc., and fundamentally restructure them with a disruptive innovation of some kind or another. If one can crack open the existing social container of economic value and strongly reconfigure its flows and contents, recapitalize its terrain, then the rewards will come pouring out, captured as profits for some new business enterprise (Clayton 1997).

Christensen's conviction is that, in fact, such disruptions are inevitable given the continuing emergence of new forms of hardware and software that eventually and challenge and destroy the status quo in just about any form of organized social activity one can mention. Disruptive innovations occur when a new product or idea "transforms an existing market or sector by introducing simplicity, convenience, accessibility, and affordability where complication and high cost are the status quo." Paradigmatic in his view was the replacement of the mainframe and min computers of earlier generations by the high powered "personal computers" of the 1980s and since.

Christensen and his followers now apply this way of thinking in many areas of business and social life, including education. His book on that topic, *Disrupting Class: How Disruptive Innovation Will Change the Way the World Learns*, makes it clear that innovation in the schools must be aggressive and forceful. At the book's conclusion he advises his readers:

[4]Marx, K., & Engles, F. *Manifesto of the communist party*, Part I. https://www.marxists.org/archive/marx/works/download/pdf/Manifesto.pdf.

> The tools of power and separation, though they seem foreign to leaders who have been schooled in consensus, are the key pieces of the puzzle of education reform. As you face budget crises and difficulty finding teachers, don't solve the problems by doing less in the existing system. Solve it by facilitating disruption. (Clayton 2008)

It's interesting that he openly embraces the classic military and political strategy of divide and conquer. As a program of restructuring and possible improvement for long-standing institutions, their commitments, practices, and practitioners, Christensen's worldview revises an old American maxim: "If it ain't broke, don't fix it." That now becomes, "If it ain't broke, by all means break it!" As if to turn the classic Hippocratic Oath for medical and professional ethics on its head, fashionable maxim seems to be: "First do some harm!"[5]

8.5 Jewel in the Crown

Today's ideas of "innovation" inherit the optimistic aura of a god term noted earlier—the idea of "Progress," a notion that long expressed the highest aspirations of modernity, hopes that looked to continuing expansion of scientific knowledge, embodied in technological advance, leading to inevitable improvement in nutrition, health, mobility, and other kinds of material well-being along with general improvement in social, moral, and political conditions. An early statement of this dream was offered by Rene Descartes in his *Discourse on Method.* "'*I perceived it to be possible to arrive at knowledge highly useful in life…to discover a practical, by means of which, knowing the force and action of fire, water, air the stars, the heavens, and all the other bodies that surround us, as distinctly as we know the various crafts of our artisans, we might also apply them in the same way to all the uses to which they are adapted, and thus render ourselves the lords and possessors of nature.*"[6]

While twenty-first century hopes for "innovation" echo beliefs of this kind, there some crucial differences. In everyday parlance "innovation" is usually regarded as a matter of limited application often geared to market-centered benefits. A product or idea is useful, accessible, flexible, visually appealing, and lower cost than the available alternatives. It helps a business or organization vanquish the competition and capture savings and profits. But a central feature of the classic idea of progress —that there is an inevitable, universal tendency toward improvement in living conditions for *all of humanity*—that is no longer part of the program. Indeed, in today's fashions of thinking and talking, "innovation" is perhaps best seen the jewel in the crown of neoliberalism, the ruling ideology of our time, a pungent worldview

---

[5]A recent survey of the sources of actual innovation finds that "destruction" is not as productive as its advocates claim. Daniel Garcia-Macia et al. How destructive is innovation? Working Paper 22953, National Bureau of Economic Research, December 2016.

[6]Descartes, R. *Discourse on method*, part IV.

that promotes economic liberalization, privatization, "free trade," open markets, deregulation, privatization of formerly public institutions, and cuts in government spending (especially for social programs) in order to enhance the role of the private sector in the economy and society as a whole. From this standpoint, market-based approaches are generally thought to be superior because they foster a spirit of innovation by those who hope to profit from their success. The core belief is that the world will improve incrementally by the proliferation of clever innovations that succeed in the global marketplace. Hence, we can celebrate the renunciation of any widely shared project to realize *the common good* because improvement will be achieved in other ways. And we can welcome the destruction of institutional and material frameworks that previously sought to realize social, economic, and political well-being as a project within the public sphere. Now secured by the efforts of clever people in their private pursuits, an enhanced, often upgraded world pours steadily from the laboratories and corporations. This is how things get better! (Eagleton-Pierce 2016).

The widely heralded features of "creative destruction" and "disruptive innovation" play a role in the episodes of "shock doctrine" described in the Naomi Klein's analysis of the excesses of neoliberal policies around the globe (Klein 2006). Klein argues that when an earthquake, tsunami, war, revolution, coup d'etat, or financial crash upsets the existing order of things, often the event is seized upon as an opportunity for radical, right-wing, market-centered restructuring. In the several case histories she analyzes, one finds deliberate, well-planned programs of shock doctrine, almost always to the benefit of political oligarchs, autocrats, billionaires, and global firms. By the same token, policies of this kind are direct alternatives to egalitarian strategies that seek "the common good" within the societies affected. Policies of neoliberalism—outsourcing, privatization, anti-unionism, deregulation in banking and environmental protection, dismantling of public services, pressures to maintain low wages, the imposition of a mountain of debt on college students, and similar measures are now widely recognized to have contributed to the steady erosion of the incomes and life prospects for the middle class and working people in the U.S.A. Thus, glowing hopes for "innovation" are all that remains as more conventional doors to social betterment are slamming shut.

8.6 Expecting Miracles

Within political discussions of the great issues facing humanity at present—world poverty, inequality, energy crisis, resource depletion, environmental ills, global climate change, etc.—the idea of innovation plays a crucial role. Many observers are inclined to say, "If only we were innovative enough, these problems would likely be solved."

Is this confidence warranted?

Will we innovate ourselves out of the rapidly widening gaps of inequality in wealth and income that afflict many world societies?

Will we innovate ourselves away from the utter dependence upon fossil fuels upon which modern civilization depends?

Will we innovate in ways that eliminate the rapidly moving threat that global climate crash poses to countless biological species, including our own?

A notable example of the widespread tendency to insert the idea of "innovation" when confronted with a world-historical crisis comes from the statements and practical commitments of Microsoft mogul Bill Gates. His vision of this strategy was briefly in a 2010 TED Talk on global warming, "Innovating to Zero."[7]

"We need solutions," Gates exclaimed, "either one or several that have unbelievable scale and unbelievable reliability….".

"These breakthroughs, we need to move those at full speed, and we can measure that in terms of companies, pilot projects, regulatory things that have been changed." Gates mused optimistically that if the expected technological miracles all happened as expected, the world could reach zero carbon emissions within the decades ahead.

To his credit, Gates has recently pledged $1 billion of his own wealth and organized a group of billionaire friends to support research and development on clean energy. Along with its anticipated role in combatting climate change, the project would also address several other global problems. "If we create the right environment for innovation," Gates proclaimed at the project's debut, "we can accelerate the pace of progress, develop and deploy new solutions, and eventually provide everyone with reliable, affordable energy that is carbon free. We can avoid the worst climate change scenarios while also lifting people out of poverty, growing food more efficiently, and saving lives by reducing pollution."[8] Among the specific steps needed in his view are increased government and private funding for research on clean energy solutions accompanied by the creation of economic incentives that can speed such "solutions" into the market.

8.7 Procrastovation

One can compare the enthusiasm Mr. Gates and other techno-optimists of our time with reports on climate change and its consequences that issue from scientific organizations published in a steady stream these days. Conveying an increasing sense of urgency, many scientists and policymakers insist that the situation is already spinning out of control and that the time to cut carbon emissions is very short. Hence, international agreements reached in 2015 at the COP21 climate change conference in Paris established a limit of 2 °C increase by the end of the

[7]Bill Gates, "Innovating to zero!" TED2010.

[8]Davenport, C. Bill Gates expected to create billion-dollar fund for clean energy. *New York Times*, November 27, 2015. http://www.nytimes.com/2015/11/28/us/politics/bill-gates-expected-to-create-billion-dollar-fund-for-clean-energy.html?_r=0.

century with an urgent goal of no more than 1.5 degrees over preindustrial levels. While these targets are a strong step forward when compared to earlier negotiations, for many observers they seem too little, too late. As Professor Stefan Rahmstorf of the Postdam Institute for Climate Impact Research warns, "These [climate change records] are very worrying signs and I think it shows we are on a crash course with the Paris targets unless we change course very, very fast. I hope people realize that global warming is not something down the road, but it is here now and affecting us now."[9]

As regards measures that could reduce carbon emissions, there are a number straightforward, readily alternatives in social policy that are often proposed: instituting stiff carbon taxes, drastically reducing energy use in housing and everyday patterns of consumer behavior, lowering automobile speed limits, drastically reducing airline travel, etc. The tendency to look toward expected technological miracles in the longer term rather than immediately deploy mundane alternatives is perhaps the primary contribution of the Cult of Innovation to today's spiraling climate crisis. A good name for this deeply entrenched obsession would be "Procrastovation"—putting off until tomorrow the practical steps that could offer substantial improvement right away.

In that light, an important but seldom asked question is this: Today's cherished rituals of Innovation/Procrastovation are an alternative to what?

While there are many conceivable answers, a good beginning would be to recognize the need for widespread democratic deliberation, inquiry, debate, and formation of a strong resolve to tackle serious problems that face the global community. A strategy of this kind was enthusiastically launched at the U.N. Rio Summit of 1992 and produced, among other accomplishments, an emphasis upon the challenge of balancing economic prosperity with measures to ensure long-term environmental well-being. Along similar lines, it seems sensible to reinvigorate widely known varieties of public planning, including plans for the deployment of technologies and institutional programs that already exist and likely to bring fruitful results.

If, as many climate scientists now insist, there is only a decade or two to achieve substantial reductions in the carbon emissions before the situation turns truly critical, can we afford to procrastovate any longer?

In fact, within the international community of those looking for effective remedies to address climate change, there is now a heated debate about pathways of deployment for existing problem-solving methods in contrast to exotic dreams for techno-magical devices anticipated on the distant horizon. Many have begun to identify the obsession with "innovation" as the last great excuse that prevents societies from taking the bold, quick, decisive steps needed to curtail the

---

[9]Quoted in Damian Carrington. Shattered records show climate change is an emergency today scientists warn. *The Guardian*, June 17, 2016. https://www.theguardian.com/environment/2016/jun/17/shattered-records-climate-change-emergency-today-scientists-warn.

civilizational bad habits that have placed Earth's biosphere in jeopardy.[10] As entrepreneur Jigar Shah observes, "Solar and wind are winning around the world not because of fundamental technological breakthroughs, but instead because after 30 years the banking sector is finally comfortable scaling up their use… I am not against innovation, but we don't need to be telling people that we need it to reach the 2 degrees milestone being talked about in Paris. What we need is deployment" (Shah 2015).

Despite warnings of this kind, a stubborn preference for "innovation" over more conventional readily available approaches to problem-solving is common among those who imagine themselves to be the sole saving remnant—philanthropic visionaries guarding humanity's future. Perhaps tendencies of this kind that such tendencies to stem from deeply ingrained occupational habits, ones that have worked so well in those in the world of business. For celebrities in Silicon Valley and other high tech meccas, "innovation" is what they know how to do, where they've been so successful. But as evidence of global poverty, inequality, energy crisis, and collapse of Earth's climate system become more difficult to deny, our technical elites often revert to a kneejerk reflex, seeing such troubles an exciting opportunity for venture capital, a chance to launch a bunch of cool new "startups."

8.8 Homilies

For the followers of powerful "god terms" and their inspiring worldviews, a set of moral lessons is included as part of the package. Its standards identify the truly worthy persons as compared to those found lacking or even obstructive when it comes to achieving the dream. Thus, for those mobilized by the myth of "Revolution" and imaginaries of "Frontier" it made perfect sense to identify the chosen people, those expected to carry the banner bravely into the future, as against the benighted souls who stood in way of positive change, groups who needed to be removed, exterminated or harshly dealt with in some way or another.

Among devotees of the beliefs and rituals discussed here, a pungent set of judgments assigns praise and blame. As long as one is being "innovative," things are bound moving on the right path and likely headed toward wonderful outcomes. "Just do your best to become one of the creative elect and the blessings will surely shower down upon you." In my experience, much of university teaching now generates homilies of that kind, reassurance for young people who spend huge sums of money in hopes of joining the clever, rich, and powerful.

Even more relevant to most of the world's seven billion people, however, the moral philosophy of innovation also contains a pungent (but often unspoken)

[10]Gaddy, B. *The innovation vs. deployment debate in energy: How did it get so heated?*. https://www.greentechmedia.com/articles/read/revisiting-the-innovation-versus-deployment-debate-where-did-it-come-from.

message of blame. "If you are not among the 'innovative' souls actively seeking to disrupt and reconfigure long-standing practices, enterprises, tools, and institutions, then you have only yourself to blame for the dreary fate that surely awaits you in the years ahead. You had your chance and you blew it! And please don't come to us asking for any sympathy or handouts. That's old school thinking—ideas about the common good, a just society and all that. We've moved beyond those dreary, outmoded concerns." One does not have to go to very many Silicon Valley workshops or seminars by smooth talking business school gurus to receive the uplifting message: "Innovate or Die! Innovate or Perish!"

In sum, "innovation" appears magically before our eyes—the jewel in the crown of neoliberalism, a fabulous gift promised those who doggedly pursue the economic and technological obsessions of the early twenty-first century. Within this aura The Cult of Innovation steadily expands—a throng of true believers engaged in ceaseless adoration of a beloved but ultimately tawdry (perhaps even deadly) treasure.

References

A recent survey of the sources of actual innovation finds that "destruction" is not as productive as its advocates claim. Daniel Garcia-Macia et al. How destructive is innovation? Working Paper 22953, National Bureau of Economic Research, December 2016.

Among the websites I visited were these: "The 25 Best Inventions of 2015," TIME Staff. http://time.com/4115398/best-inventions-2015/. "The 100 Greatest Innovations of 2015," Popular Science Staff, October 23, 2015. http://www.popsci.com/100-greatest-innovations-2015.

Center for Social Innovation, Stanford Graduate School of Business. https://www.gsb.stanford.edu/faculty-research/centers-initiatives/csi.

Clayton, C. M. (1997). *The innovator's dilemma: When new technologies cause great firms to fail*. Cambridge: MA, Harvard Business School Press.

Descartes, R. *Discourse on method*, part IV.

Clayton, C. M. (2008). *Disrupting class: How disruptive innovation will change the way the world learns* (pp. 226–227). New York: McGraw Hill.

Davenport, C. Bill Gates expected to create billion-dollar fund for clean energy. *New York Times*, November 27, 2015. http://www.nytimes.com/2015/11/28/us/politics/bill-gates-expected-to-create-billion-dollar-fund-for-clean-energy.html?_r=0.

Gaddy, B. *The innovation vs. deployment debate in energy: How did it get so heated?*. https://www.greentechmedia.com/articles/read/revisiting-the-innovation-versus-deployment-debate-where-did-it-come-from.

Eagleton-Pierce, M. (2016). *Neoliberalism: The key concepts*. New York: Routledge.

Gates, B. Innovating to zero! TED2010. https://www.ted.com/talks/bill_gates.

Internet of Caring Things, April 2014, Trend Briefing. http://trendwatching.com/trends/internet-of-caring-things/.

Joseph, A. S. (1950). *Capitalism, socialism and democracy* (3rd ed., pp. 81–82). New York, N.Y.: Harper Torchbooks (1962).

Marx, K., & Engles, F. *Manifesto of the communist party*, Part I. https://www.marxists.org/archive/marx/works/download/pdf/Manifesto.pdf.

Klein, Naomi. (2006). *The shock doctrine: The rise of disaster capitalism*. New York: Metropolitan Books.

Quoted in Damian Carrington. Shattered records show climate change is an emergency today scientists warn. *The Guardian*, June 17, 2016. https://www.theguardian.com/environment/2016/jun/17/shattered-records-climate-change-emergency-today-scientists-warn.

Richard, M. (1953). *Weaver, the ethics of rhetoric*. Davis, CA: Hermagoras Press.

Shah, J. (2015). *Jigar shah to Bill Gates: We already have the technology to solve climate change*. ImpactAlpha, December 2, 2015. http://impactalpha.com/dear-bill-gates-we-already-have-the-technology-to-solve-climate-change/.

Chapter 9
A Generative Perspective on Engineering: Why the Destructive Force of Artifacts Is Immune to Politics

Ron Eglash

Engineering is closely tied to social and environmental destruction across the globe. The energy industry has created global warming, oil spills, acid rain, and toxins ranging from mercury to radioactive waste. Manufacturing has turned skilled crafts into low paid, repetitive assembly jobs. Information technology has accelerated wealth inequality such that the top 1% of the world's wealthy now own 50% of the wealth. One theory is that these detrimental effects have nothing to do with the engineering design; that they are merely the result of how capitalism forces us to use otherwise "neutral" technologies. But socialist experiments from the USSR to Venezuela have shown the same degree of pollution and poverty as capitalism. "The People's" radioactive waste left over from the USSR will kill you just as fast as General Electric's radioactive waste in the US. *The destructive force of artifacts is immune to politics*. However that is because both left and right ends of the political spectrum have focused on systems built for the extraction of value. A generative approach to engineering, in contrast, would design technologies specifically for maintaining value in unalienated forms, and circulating that value rather than extracting it. This paper will review this underlying concept of generative justice, and how that can be adapted to engineering practice.

9.1 Labor Value, Ecological Value, and Expressive Value

Elsewhere (Eglash and Garvey 2014; Kuhn 2016; Eglash 2016a, b), we have explained generative justice by starting with examples of indigenous cultures. In the traditional communitarian economy of the Iroquois for example, an

R. Eglash (✉)
Department of Science and Technology Studies, Rensselaer Polytechnic Institute (RPI), Troy, NY, USA
e-mail: eglash@rpi.edu

E. Subrahmanian et al. (eds.), *Engineering a Better Future*,
https://doi.org/10.1007/978-3-319-91134-2_9

"agroecology"—the three sisters of corn, beans, and squash—had dramatically higher yields than European plow methods of the day (Mt Pleasant 2006, 2011). Bean root nodules contain nitrogen-fixing bacteria that help corn and squash; corn provides vertical support for beans; squash's broad, spiny leaves prevent soil moisture loss, weeds, and pests. The soil agroecosystem was so effective that Euro-American farmers who annexed their land in 1804 reported initial yields of 80 bushels per acre. By 1845 yields had dropped to 26 bushels per acre; without the Iroquois ecosystem, the soil was rapidly depleted. Even today, artificial pesticides and chemical fertilizers fail to achieve the long-term effects that agroecosystems make possible (creating, for example, pesticide "treadmills").

It is easy to understand, in the case of soil depletion, what is meant by "value extraction". A flow of things that have value with respect to ecosystem flourishing —nutrients, physical features, hydration—move in a cycle. Human harvests by the Iroquois were part of that cycle. The Euro-American farmers extracted value rather than cycling it. Once extracted and sold for cash, we can say the value was "alienated"—converted to a form the ecosystem cannot use. Other examples of ecological value extraction include overfishing, unsustainable logging, and so on.

The concept of "value alienation" is most closely associated not with ecological value, but rather with Karl Marx's analysis of labor value. In his 1844 "comment on James Mill", he asks us to imagine a traditional artisan whose pride in excellent crafting skills and pleasant social ties in a precapitalist village provide the greatest satisfaction in life; here "the act of labor itself is for him the enjoyment of his personality and the realization of his natural abilities and spiritual aims." Marx later used the Iroquois specifically as an example, since Lewis Morgan had documented how their relations of reciprocity, communal sharing, and gift-giving ensured unalienated labor value circulation.

Marx contrasts that vision of traditional artisanal satisfaction with the physical, financial, and psychological deprivation of low-skilled workers in industrial factories: the worker has become alienated from their product (one cannot take pride in having repeatedly inserted screw #17 on the assembly line all day); alienated from their work process (see "assembly line"); relations with users (who might not buy the product if they could see the suffering attached to it); and even their own bodies (e.g., repetitive strain and industrial toxins). From Marx's point of view, the labor value that one could have invested in meaningful artisanal work has now become alienated from the worker.

Although Marx was primarily focused on labor value, some of his remarks on soil depletion show that he was aware that ecological value was also circulated in traditional societies. And although he had no category for it, he occasionally mentions what I would place in a 3rd category, that of "expressive value". The Iroquois, for example, had voting for women centuries before any European nation. Like most Native American cultures, they provided a legitimate role ("Two Spirits") for people we would consider gay, lesbian, bisexual, and gender-variant. And while neither Marx nor the Iroquois might recognize them all, most of the things we consider protected by civil rights in our era—free speech, the right to be an atheist or practice a religion of your choosing, free access to knowledge, love for people and places, and so on—would also be examples of expressive value. Like labor value and ecological value, the generation of expressive value also best flourishes when it is freely circulated, and can be extracted to the detriment of those who generate it (think, for example of the ways religious faith is extracted for political gain).

In all three cases—labor value, ecological value, and expressive value—the promise that capital makes ("no worries, we will return that value back to you in the form of money") is a false one, because once systems of work are designed to maximize the extraction of value—mass production, deskilling, "externalizing costs" such as health and environment protections—the damage has already been done, and buying commodities to compensate merely immerses us further in alienated products.

Marx thought that taking capitalism out of the equation would solve the problem, but he was mistaken. One of the best sources for this comparison is sociologist Burawoy (1985), who carried out participant observation studies as a factory worker in the manufacturing industry in Chicago, and similar plants in communist Hungary and the USSR. As a life-long dedicated Marxist, he had no personal inclination towards reporting the negative side of state socialism, so his critiques are all the more convincing. In both capitalist and communist industries he found similar deprivations—dangerous environments with limited safeguards; low pay; and his main focus, forms of coercion that keep people working hard. The methods of coercion were different but equally damaging: "each system has its own rationalities and irrationalities, and each fashions workers who adapt to or resist those (ir)rationalities" (Burawoy 2006, p. 65).

Rather than extract value and centralize it for later redistribution, it is possible to have a generative economy: leave value in unalienated form, and circulate it through a commons. Hence the definition of generative justice (Eglash 2016a): *The universal right to generate unalienated value and directly participate in its benefits; the rights of value generators to create their own conditions of production; and the rights of communities of value generation to nurture self-sustaining paths for its circulation.*

Marx thought that extraction and centralization would be required for high-tech societies; the generative ideal would only be possible for low-tech indigenous societies. But generative cycles are indeed possible in high-tech circumstances as well: open source computing is a common example. The challenge is that since we are starting from an extractive economy, it is hard to kick-start an entirely new mode of living: for example if you give away code for free, how do you make a living?

In Eglash (2016b), we provide an example of such a transition in the case of Arduino, an open source microprocessor (Fig. 9.1). We have visualized the flow of value in two ways. When alienated it appears as a single line; when unalienated as double lines. In the upper left quadrant, we see mass production of computer chips as the extraction and alienation of labor value as Marx envisioned it: low-income workers with little benefits or pay. In the lower right we see unalienated value flow without income; the Internet "gift economy" of makers. But in the upper right, we see a hybrid cycle in which both the gift economy and realistic income converge.

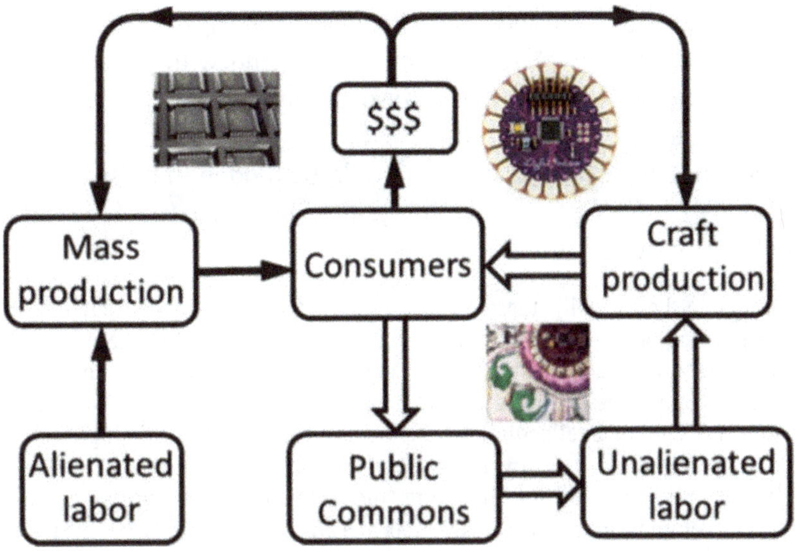

Fig. 9.1 The links between extractive and generative flow in the Arduino ecosystem

This example is a modified Arduino board, the circular LillyPad created by Leah Buechly to reduce gender barriers to DIY electronics and computing. As a for-profit company, it is creating income. But as part of the gift economy, its design and code is open source, and users freely give away their designs to a commons, from which they too benefit. In Eglash (2016b), I have detailed many examples of such beneficial social relations embedded in the Arduino ecosystem—in particular, social justice and environmental sustainability projects—and the ways that have enabled other instances of this "third cycle" in which a gift economy is linked to forms of "social entrepreneurship".

The Arduino ecosystem is not an independent generative economy. It is still tied to exploitation of labor and nature in some of its electrical components in the upper left quadrant, and even the other two are still vulnerable to problems such as sexism in makerspaces (Dunbar-Hester 2016). But it creates a clear goal for the transition to generative justice: shrink the upper left quadrant and expand the two on the right. What kinds of engineering practices can contribute to that pathway? What kinds of engineering opportunities do we need to be attentive to when asking how we can replicate such cycles elsewhere? The next section will examine both historical and contemporary case studies.

9.2 Watchmaking

There have always been generative alternatives to extractive forms, and they do not all lie in a low-tech indigenous past. For example, in 1872 Russian scientist Peter Kropotkin traveled to the Jura Mountains in Switzerland. The communities in the region were famous for having defended their industry against corporate takeover by the pressures of mass production, which was turning out cheaper (and lower quality) products elsewhere. And yet they also defended the autonomy of groups within the International Workingmen's Association (IWA), rejecting an attempt by Karl Marx and his followers to turn the IWA general counsel into the central authority of a political party. What power could be hidden in these small towns that could withstand the pressures of both right wing and left wing authoritarians? Kropotkin (1899) writes:

> In a little valley in the Jura hills there is a succession of small towns and villages, of which the French-speaking population was at that time entirely employed in the various branches of watchmaking. …The very organization of the watch trade, which permits men to know one another thoroughly and to work in their own houses, where they are free to talk, explains why the level of intellectual development in this population is higher than that of workers who spend all their life from early childhood in the factories. … The egalitarian relations which I found in the Jura Mountains, the independence of thought and expression which I saw developing in the workers… appealed far more strongly to my feelings; and when I came away from the mountains, after a week's stay with the watchmakers, my views upon socialism were settled. I was an anarchist.

I selected the Jura example because this volume is dedicated to a conversation between engineering and social science. Illustrating the concept of unalienated value with indigenous culture seems like something that appeals best to anthropologists, but I hope that watchmaking is an example of unalienated value can speak to engineers: the material and intellectual demands for precision, metallurgy and mechanics; what Csikszentmihalyi (2000) refers to as the mental state of "flow" during a skillful crafting experience; pride in design innovation; and respect for independent, rational thinking. Beyond individuals, the collective social dimension here is quite significant. Kropotkin immediately saw the critical role that was played by workers in charge of their own production environment; in particular the role of "expressive value": free speech and inquiry, intergenerational relations of caring (Folbre 2014), and other features were cycled within this network as well. Veyrassat (1997) compares the success of Jura watchmakers to the failure of the calico-printing industry in yet another Alpine valley in Switzerland during this same time period. She shows that the calico printers attempted to preserve wages and working conditions by passing laws against innovations such as "double printing"; in doing so they became vulnerable to advances elsewhere. In contrast "the watch industry was to set out on the path to a modernization that did not break with the indigenous manufacturing model" (p. 201).

Auerswald (2017) points out that the Jura watch tradition did not stop there: In 2014 China exported 669 million watches; 20 times that of Switzerland thanks to the role of automation and robotics. Yet robot profits did not exceed those of highly skilled humans: At $24.3 billion, the Swiss watch industry made 5 times that of the Chinese. On the other hand, contemporary Swiss watchmaking is no longer the province of working class artisans. Auerswald is not insensitive to this issue; he points out that the American company Shinola, located in Detroit, imports Swiss watch parts and assembles them into American-branded casing. He cites this example—a new company birthed in the very city that symbolizes American manufacturing decline—as vindication for his thesis that the increasing abstraction of technology—the tendency to move from physical mechanism to code—spontaneously creates new entrepreneurial niches, due to "an inexorable evolutionary logic that constantly shifts the landscape of opportunity".

What Auerswald fails to note is that Shinola's location in Detroit was part of a carefully calculated marketing strategy. Muller (2013) describes how Texas billionaire Tom Kartsotis first did marketing research. He discovered that a luxury item branded as "made in Detroit" made the product attractive enough to compete against luxury imports; he then purchased the Shinola name from the defunct shoe shine company to add a nostalgic aura. As Perman (2016) puts it,

> With Shinola, Kartsotis has performed a near magical marketing act–creating an artificial heritage brand by co-opting others' rich American histories. …Shinola's products are designed and packaged with an American midcentury look, evoking nostalgia for a bygone era of quality and integrity. Most important, by hatching the brand in Detroit–a city emblematic of American hardship, resilience, and craftsmanship–the brand is selling more than watches; it's selling a comeback. Every time customers in Neiman Marcus or Saks purchase one of the brand's $850 watches or $300 leather iPad cases, they too can feel like they're doing their part in Detroit's fight for survival.

In other words, Auerswald has it backwards. It is not an "inexorable evolutionary logic" of progressive technological abstraction that spontaneously created a new job niche, and they just happened to locate it in Detroit. It is because Kartsotis and his market research discovered they could tap into a yet-to-be-exploited source of expressive value: the human desire to love our cities despite their decay; to root for the underdog; to live a morally acceptable life. And our human fears too: our feeling that as rich Americans wearing a luxury European brand we might be seen as traitors, while wearing an American brand—even one made from Swiss parts—will make us feel good about ourselves, because we see ourselves through the eyes of others.

Using the colloquial term "bougie" [boo-zhee] for bourgeois, professor of design Modrak's (2015) insightful essay titled "bougie crap" examines the contradiction between the working class cultural capital that Detroit represents, and the lack of return value flow to the working class in the case of Shinola. She attributes that in part due to the product itself (hence the title), and in part the resulting gentrification:

> Start with a neighborhood or city that lacks economic incentives or that is populated by minority groups, which are underserved by municipal services including education, transportation, street lighting, police response time and maintenance. Enter a mainly white, middle-class population. Investors clamor to underwrite new businesses, sponsor grants or to secure real estate. This triggers a spike in real estate prices and a flood of new commercial ventures that sell expensive bougie crap that only the new residents can afford.

I do not mean to be dismissive about the gamble Kartsotis made in locating in Detroit, or even the idea of tapping into such wellsprings of expressive value. Rather, we should focus on the missed opportunity to return value to those who generated it. As a counterexample, consider Bachinger's (2015) analysis of the

VinziRast coffee cooperative in Vienna (Fig. 9.2). Just as Detroit has a symbolic heritage in its manufacturing history, Vienna has one in its coffee houses.

They were the historic hangout for intellectuals, artists, and activists from Sigmund Freud to Leon Trotsky; the target of Nazi closures in 1938; and today officially designated as "intangible cultural heritage" by UNESCO. Like Shinola, VinziRast taps into this flow of expressive value rooted in a civic history. But VinziRast is not a relationship of value extraction: All the café profits go to an NGO, and the business is part of an innovative housing project where students and formerly homeless people live, learn, and work together. The café employees are drawn from this low-income population, the food is locally grown, and even the supply chain transportation is sustainable; using bikes they have modified with loading bins.

As in the case of Arduino, the flowchart shows three linked cycles: an exploitative relationship in the upper left (the image shows child labor on a Nicaragua coffee farm); a nonprofit cycle of commons-based value flow in the lower right; and the hybrid cycle of VinziRast in the upper right. Using a sliding scale for its café prices, VinziRast allows consumers to democratically decide how to modulate the links between gift economy and profit economy. Again, the question is how we can expand the right side's generative cycles, and diminish the left quadrant's extraction. Shinola's expanding profits and product line are empowered by technological and design innovation; could that be adapted by VinziRast? Conversely, could VinziRast's cycle of unalienated value flow (assuming the political and financial will to do so) be adopted by Shinola? Or is there something inherent in engineering technology that locks these two on opposite sides of a divide?

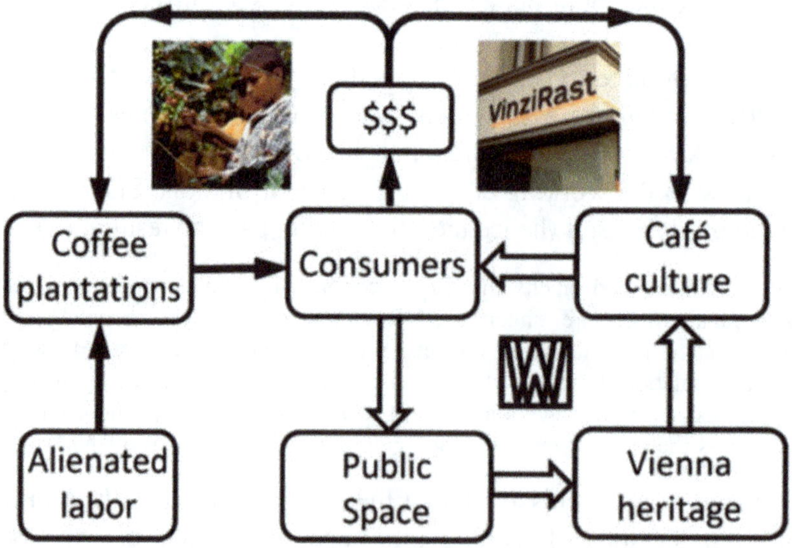

Fig. 9.2 Generative and extractive cycles in the VinziRast coffee house in Vienna. Coffee farm photo courtesy of Trocaire

I find that such possibilities for synthesis are too often answered with relatively minor tweaks: perhaps Shinola offers to add donations to some local charity, or VinziRast adds a cell phone app for ordering. Haraway's (1991) cyborg metaphor is a useful starting point for dismantling the barriers that keep this divide in place. She notes that creating a divide in which social justice and sustainability are pushed to an organic, low-tech side, and naming "the enemy" to be everything on the other side ends up reproducing many of the misleading assumptions that caused the problems in the first place. Authoritarian claims for "the natural" are common in the history of injustice: when LGBTQ people are accused of "unnatural sex" or interracial relationships are seen as violating national purity and its union of "blood and soil" we witness the negative consequences of the natural/artificial dualism. Instead Haraway urges us to recognize that humans are "always already" part artificial: our jaws evolved to their present tiny size because we invented fire; our immune systems have been reprogrammed to ward off polio, diphtheria, and other deadly diseases; everything deeply human about us, from language to clothing to shelter, draws as much from the artifice of innovation as it does from nature. That is not to say that making something "more cyborg" automatically makes it more just or sustainable; rather it is a call for considering multiple paths that do not exclude the cyborg options.

9.3 A Cyborg Path to Generative Justice

One of the pressing issues of our historical moment is the dramatic increase in automation, from AI to robotics. Auerswald's above discussion of watchmaking was intended to address that issue in two ways: first, he claims that tasks which tend to be more human-centric (in this case high-end luxury watch crafting, but he also mentions waitress, cook, actor, etc.) will always provide a safe economic refuge, even against automation challenges. Second, he claims that automations' encroachment spontaneously creates new economic niches, and some of those will be suitable for exactly those human-centric tasks. I hope the description of Shinola's marketing strategy above provides a useful counter to the second claim: markets are strategically created, not simply handed to us by inexorable techno-evolution, and the available strategies are increasingly in the hands of the already-wealthy. As to the first claim, I will offer a counterargument from Noble's (1986) analysis of General Electric's (GE) experiments in replacing skilled machinists with numerically controlled (NC) machine tools—automated devices that shaped metal according to a computer program—during the early 1960s.

Nobel shows that the tools were not the simple result of superior technology replacing inferior humans: the machinists' hand-guided product was, during the initial years, superior to that of the NC automation in both quality and quantity. Rather it was a deliberate strategy to break labor union shop floor control, and enhance "Taylorism" in which workers behavior is strictly controlled from the top down. Putting computer programmers in charge of NC tools was an attempt to put shop floor production in the hands of a white collar task (coding), a natural alliance with management, leaving the machinists as deskilled machine button pushers. But an important objection to Noble's account was raised by Pickering (1995).

Picking begins by citing Haraway, noting that the machinists and their tools already formed a "cyborg" pairing prior to NC automation. In his language, when human agency interacts with machine agency they always form a "mangle" of the two, and any destabilization—introducing new technologies, social formations, etc.—will always contain an element of unpredictability; a "dance of resistance and accommodation" as human intentions and nonhuman forces negotiate until re-stabilization occurs. He notes that Noble's own analysis—the Marxist claim that class domination will always be the overriding force, even above profits or productivity—failed to predict what happened next. Once the poor quality of NC tool production was made apparent, GE management attempted the usual set of rewards and punishments to fix the problem. When that failed, they made a desperate move: a pilot program put these workers in charge of production.

A radical experiment in worker control given the context, GE's new Pilot Program would be "unique in that there was to be no foreman, no scheduled lunch periods, and flexible starting and personal times." (Nobel 281). The generative opportunity was not lost on workers: machinists began to "schedule equipment start-up; work with planning in developing, implementing, and controlling new methods and procedures; approve programming from the viewpoint of good machine shop practice; review and make suggestions about changes in workstations, tools, and fixtures; assume responsibility for quality in the unit and interface with quality control; [and] monitor the area for availability of all materials and check equipment to insure safe and proper functioning" (280–281). Production and quality dramatically increased. As newly empowered workers began to clash with management, GE administration concluded that the risks were now exceeding their benefits, and enough had been learned to reestablish the older managerial style.

From Pickering's view, this shows that his framework of the emergent human–machine "mangle" offers a better understanding, because the worker-controlled production, even if only temporary, was unpredicted. From Noble's view, the fact that managerial dominance was eventually reasserted proves his Marxist framework; the momentary contradiction was merely a sham for duping workers into exploiting themselves. But I want to draw out a third possibility by comparing the GE pilot program to a strikingly similar event happening at exactly the same time period on the other side of the iron curtain. In 1968, many factories in Czechoslovakia also began to experiment with worker control.

The backdrop of the Czech experiment is always described in political terms, and for good reason. At the end of WWII, Czechoslovakia was the only nation that voluntarily voted to be on the Soviet side, and they naively thought that they could create their own vision for communism. The "Prague Spring" of 1968 saw a brief liberalization in many domains: fewer travel restrictions, less censorship, more consumer goods. The brief rise of worker-controlled factories is usually seen as a purely political outcome of this movement. But there was also an undercurrent of the same internal engineering critiques that plagued GE regarding parts quality.

The communist version of Taylorism (in the USSR promoted by *nauchnaia organizatsia truda*, “the movement for the scientific organization of labor”) was similar to GE’s top–down control methods. In a 1968 speech promoting the worker self-management movement, one of the engineers described a similar disappointment in the quality of machined parts:

> When we steel workers pointed out that we were turning out steel for the scrap heap, they nearly put us in jail because, they said, we were throwing mud at our socialist industry. … Aren’t you ashamed,’ they said ‘you’re steel workers and you criticize the steel concept. You’re reactionaries.’ Only I don’t have to be dumb just because I’m a steel worker. (Vitak 1971, p. 251)

The generative alternative for both GE factory workers and Czechoslovakian factory workers ended by different methods: at GE, they simply reasserted the old top-down management, whereas in Czechoslovakia, half a million Warsaw Pact troops and tanks invaded the county. But I hope readers can also see the underlying similarities (Fig. 9.3). Whereas Noble’s Marxist analysis positions the generative moment as an all-too predictable capitalist ploy, and Pickering sees it as evidence of the inherently unpredictable emergence, a generative perspective would view it in terms of movements along a dimension *orthogonal* to the left/right political spectrum. The question is then how to nurture the movements along that vector, adapting to the unavoidable contingencies.

9.4 Tuning for Generative Justice

“Tuning” is Pickering’s term for the series of adjustments that occur as human and machine agencies re-stabilization. I think Pickering is right about the unpredictability engendered by such a “dance of agency,” but a generative perspective

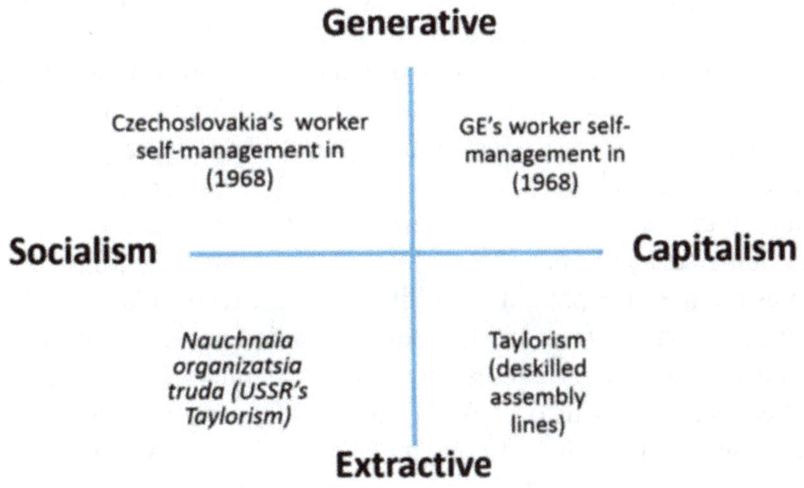

Fig. 9.3 Generative justice as orthogonal to the left/right political spectrum

can adopt that as a positive attribute; a means of exploring the potential design space. Our experiments in this area are often with indigenous groups, and when introducing our design process I frequently use the metaphor of plant roots and water. When water enters soil it undergoes percolation, trickling through whatever grains of sand or crevices offer the path of least resistance. Plant roots take a similar search, growing along whatever hints they can sense in the earth. Eventually, water and roots meet up.

In similar ways, we try to avoid the stereotype of "clever engineers here to solve your problem," or the reverse trope of fragile indigenous cultures who will be tainted by any change. Rather, each side needs to explore the space of possibilities, find meeting points where root and water intersect, and investigate the potential consequences. Reducing indigenous knowledge to a western translation is not sufficient; a recursive transformation is required (Lachney et al. 2016a). We need concepts that become <u>more</u> compelling with advancing knowledge, not less so; but even something like Haraway's "sympoiesis" cannot stand on its own. We to help engineers tune their work to the self-generating key of life, and that would not happen in the tone-deaf training camps we call STEM education.

STEM education is increasingly turning to bottom-up structures, and yet these often fail to offer generative justice because they ignore value alienation. Consider, for example, MIT's Scratch program in which children learn coding to create their own games and animations. Our recent study (Lachney et al. 2016b) showed 2960 occurrences for "Barbie," 6530 results for "McDonalds"; 4600 for "Disney Princess"; 8210 for Transformers; "17,400" results for Call of Duty; and over 3 million search hits for "Pokemon." The Scratch website's motto: "We turn children from consumers into producers." Marketing and media has colonized children's lives, For the last decade, our group at RPI has been creating alternative forms of STEM education using simulations of Navajo weaving, cornrow braiding, urban graffiti, and other practices that can empower children with their own heritage, rather than the lesson that both science and art all come from the colonized world (www.csdt.rpi.edu). As these simulations get picked up elsewhere, we have learned more about cycling unalienated value back to its source of generation, which I will illustrate with the following example from our work in Ghana (Babbitt et al. 2015).

When I said we needed concepts that become more compelling with advancing knowledge, and not less so, I did not mean to suggest that atheism is better than spiritual practices; rather it is a question of contextualizing these guiding principles in ways that advance both Technoscience and indigenous partners. In our work to establish an indigenous basis for engineering in Africa, we began with adinkra, an indigenous stamped textile practice that uses symbols representing spiritual concepts such as the lifeforce present in all living things, reconciliation with enemies, etc. Adinkra ink is made from the bark of the Badie tree, and areas in which the bark is extracted suffer less deforestation. Our first intervention was using solar energy rather than firewood to cook the ink. The second was using the adinkra symbols themselves in math and computing education; combining virtual forms with a hands-on practice that helps increase employment for ink makers and symbol carvers. The virtual form was burned to CDs for local sale. Adinkra was also used for HIV prevention. We began with surveys that indicated embarrassment at point of sale was a barrier for condom use. The inspired a project in which New York and Kumasi mechanical engineering students collaborated on the development of a locally produced condom vending machine. Adinkra artisans created an exterior to add local aesthetics. And an e-waste "upcycling" program was introduced to supply parts for both the condom machine and the computing education program.

Figure 9.4 shows the flows of value described above. Some of the engineering was quite sophisticated; for example, the local team in Ghana wanted to add 3D printing for the gears in the vending machine using recycled plastic. They are now adapting the design to fit pregnancy tests and reproductive health kits. A pessimist might assume such approaches are doomed to be restricted to small-scale enterprises or remote villages. To provide a counter example, we can look at how our indigenous simulations have been taken up by architects attempting to improve the environmental and social characteristics of large-scale buildings (Fig. 9.5).

At the same time, little of this was intuitive for the engineers involved. We need to stop training engineers to ask "what do people want"—an answer which will be conditioned by their training in capitalist extraction of value and colonization, as we saw in Scratch—but rather training them how to research and recognize unalienated value; to engineer solutions in which that value can be nurtured and circulated; and develop systems that put those who generate the value—humans and nonhumans alike—in charge of its production.

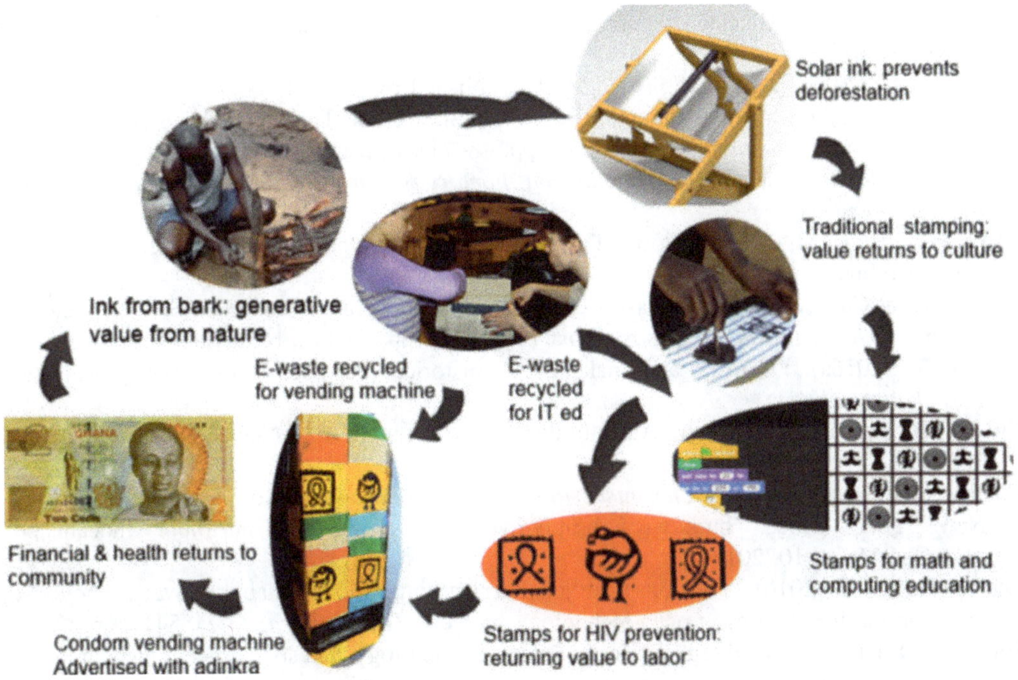

Fig. 9.4 Unalienated value flow in engineering projects in Ghana

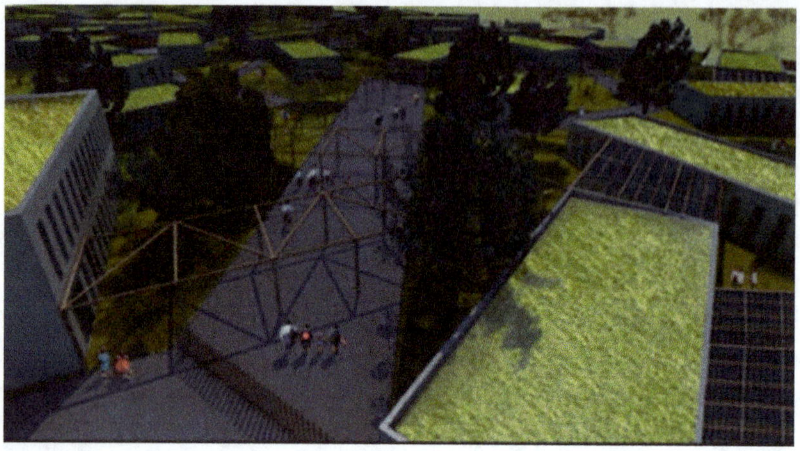

Fig. 9.5 Indigenous fractals in contemporary architecture (from http://blog.ted.com/architecture-infused-with-fractals-ron-eglash-and-xavier-vilalta/)

Acknowledgements The author would like to acknowledge NSF grants DRL-1640014 and DGE-0947980 in support of this work.

References

Auerswald, P. (2017). *The code economy: A forty-thousand year history*. Oxford University Press.

Babbitt, W., Lachney, M., Bulley, E., & Eglash, R. (2015). Adinkra mathematics: A study of ethnocomputing in Ghana. *Multidisciplinary Journal of Educational Research, 5*(2), 110–35 (June 15, 2015).

Bachinger, L. (2015). *More than just coffee*. Unpublished manuscript.

Burawoy, M. (2006). Antinomian marxist. In A. Sica & S. Turner (Eds.), *The disobedient generation: Social theorists in the sixties* (pp. 48–71). Chicago: University of Chicago Press.

Burawoy, M. (1985). *The politics of production: Factory regimes under capitalism and socialism*. London: Verso.

Csikszentmihalyi, M. (2000) Flow. The Psychology of Optimal Experience. New York: HarperCollins

Eglash, R., & Garvey, C. (2014). Basins of attraction for generative justice. In S. Banerjee, Ş. Ş. Erçetin, & A. Tekin (Eds.), *Chaos theory in politics* (pp. 75–88). Netherlands: Springer.

Eglash, R. (2016a). Of marx and makers: An historical perspective on generative justice. *Teknokultura, 13*(1).

Eglash, R. (2016b, December). An introduction to generative justice. *Teknokultura, 13*(2), 369–404.

Folbre, N. (2014). *Who cares? A feminist critique of the care economy*. Rosa Luxemburg Stiftung–New York Office. http://www.rosalux-nyc.org/wp-content/files_mf/folbre_whocares.pdf. Accessed March 10, 2017.

Dunbar-Hester, C. (2016). "Freedom from jobs" or learning to love to labor? Diversity advocacy and working imaginaries in open technology projects. *Teknokultura, 13*(2), 541–566.

Haraway, D. J. (1991). A cyborg manifesto: Science, technology, and socialist-feminism in the late twentieth century. In Simians, cyborgs and women: The reinvention of nature (pp. 149–81). New York: Routledge.

Kropotkin, P. (1899). *Memoirs of a revolutionist*. Boston: Houghton Mifflin.

Kuhn, S. (2016). Fiber Arts and Generative Justice. *Teknokultura, 13*(2), 461–489.

Lachney, M., Babbitt, W., & Eglash, V. (2016a). Software design in the construction genre' of learning technology: Content aware versus content agnostic. *Computational Culture, 5.* http://computationalculture.net/article/software-design-in-the-construction-genre-of-learning-technology-content-aware-versus-content-agnostic.

Lachney, M., Bennett, A, Appiah, J., & Eglash R. (2016b). Culturally responsive computing as brokerage: Toward asset building with education-based social movements. *Learning, Media and Technology*, 1–20.

Modrak, R. (2015). Bougie crap. *Infinite Mile 14: February 2015.* http://infinitemiledetroit.com/Bougie_Crap_Art,_Design_and_Gentrification.html.Retrieved10. March 2017.

Mt. Pleasant, J. (2006). The science behind the three sisters mound system: An agronomic assessment of an indigenous agricultural system in the northeast. In J. E. Staller, R. H. Tykot, & B. F. Benz (Eds.), *Histories of maize: Multidisciplinary approaches to the prehistory, linguistics, biogeography, domestication, and evolution of maize* (pp. 529–537). Amsterdam: Academic Press.

Muller, J. (2013). In Bankrupt detroit, shinola puts its faith in american manufacturing. *Forbes*. Retrieved March 10, 2017. http://www.forbes.com/sites/joannmuller/2013/07/26/in-bankrupt-detroit-shinola-puts-its-faith-in-american-manufacturing/2.

Noble, D. (1986). *Forces of production: A social history of industrial automation*. New York: Oxford University Press.

Pickering, A. (1995). *The mangle of practice: Time, agency, and science*. Chicago: University of Chicago Press.

Pleasant, J. (2011). The paradox of plows and productivity: An agronomic comparison of cereal grain production under iroquois hoe culture and european plow culture in the seventeenth and eighteenth centuries. *Agricultural History, 85*(4), 460–492.

Perman, S. (2016). The Real History of Shinola, America's Most Authentic Fake Brand. *Inc.com*. Retrieved May 31, 2016 from https://www.inc.com/magazine/201604/stacy-perman/shinola-watch-historymanufacturing-heritage-brand.html.

Veyrassat, B. (1997). Manufacturing flexibility in nineteenth-century Switzerland: Social and institutional foundations of decline and revival in calico-printing and watchmaking. In D. Sabel & J. Zeitlin (Eds.), *World of possibilities: Flexibility and mass production in western industrialization* (pp. 188–237). Cambridge: Cambridge University Press.

Vitak, R. (1971). Workers' control: The Czechoslovak experience. *Socialist Register, 1971,* 245–258.

Chapter 10
Does Law Wear Out?

David Howarth

The theme of *Law as Engineering: Thinking about what lawyers do* (Howarth 2014), a theme ultimately derived from Herbert Simon's *The Sciences of the Artificial* (Simon 1996), is that lawyers are in the business of creating artificial social structure. Long-term commercial contracts, wills, statutes, constitutions and international treaties are all attempts to alter the course of social events in ways desired by their designers in response to the desires of their clients. Litigation, which forms the basis of the public stereotype of what lawyers do, is a minority activity akin to military engineering. For the most part, lawyers, like engineers, spend their time not in conflictual activities but in cooperative ones, trying to clarify what their clients want (and do not want) and coming up with designs that fulfil their clients' objectives at reasonable cost. Because lawyers face design problems similar to those faced by engineers, albeit ones mainly in the social world rather than mainly in the physical, it is not surprising that their approaches overlap and that they can learn from one another.

10.1 Extended Action Through Time

One feature of lawyers' attempts at changing the course of social events is that they are usually not one-off interventions intended to push a single event in a new direction one time only. Instead, they attempt to set up constant pressure to shape events in desired ways over extended periods of time. Long-term contracts are designed to sustain a relationship between the parties, inducing all of them to deliver on their promises and resolving disputes as they arise (Deakin and Koukiadaki 2009). Similarly, statutes aim to change social behaviour, deterring

D. Howarth (✉)
University of Cambridge, Cambridge, UK
e-mail: drh20@cam.ac.uk

E. Subrahmanian et al. (eds.), *Engineering a Better Future*,
https://doi.org/10.1007/978-3-319-91134-2_10

some behaviours and encouraging or facilitating others, not just once but over and over again. Engineers too often design structures and devices intended to produce the same effect repeatedly over time—indeed most products other than food, soap, fuel and explosives have this characteristic. This suggests a project of asking whether lawyers might benefit from engineers' insights about the design of structures and devices that have this temporal component.

The aspect of the project taken up here, in a very preliminary sketch, is perhaps the most basic. Devices designed to operate through extended periods of time tend to deteriorate and eventually to fail. Physical devices wear out. Software becomes obsolete (Verma et al. 2016). Engineers have spent much effort developing methods of anticipating and correcting such deterioration. For their part, lawyers know that contracts are often renegotiated and statutes are often amended. That looks very much like deterioration over time being corrected.

More specifically, we can identify situations in which a legal device no longer produces the results desired by clients and needs to be changed to restore it to functionality. Two types of processes immediately spring to mind through which a legal device no longer produces desired results.

- The first is where different results come to be desired. One can imagine, for example, new legislators being elected who want outcomes different from those their predecessors wanted or new managers taking over a corporation who have a different risk profile. These situations are interesting from a design point of view because the designer faces a choice between starting again and attempting to repurpose an existing device. The former is expensive but the latter might not work.
- The second way a device might no longer bring about desired effects is that, without any change in what is desired, something happens to change the device's output from the desired path. For example, court decisions might alter the meaning or effectiveness of contractual terms or of statutory provisions. In these situations, the question is not how to adapt a device to new purposes but whether to repair it so that it carries out its original purpose, and if so, how to do so (a process that in practice can be messy and sticky: see, e.g. Gulati and Scott 2013).

Both processes are familiar to engineers. Here I concentrate on the latter process, but lawyers and engineers should also be able to exchange ideas and practices about the former.

10.2 A Legal Bathtub Curve?

An obvious place to start is with reliability engineering and to think about the relevant sources of failure for a legal device such as a statute (a very similar exercise could be carried out for other devices, such as treaties and commercial contracts). For physical products, we might point to 'infant mortality' failure, random failures and failures consequent on a product wearing out (Nash 2016; Verma et al. 2016). But what are the sources of legal failure? We might point to a similar trio of sources.

- First, devices might fail properly to capture clients' objectives in themselves. A tax statute, for example, might turn out to have an unintended effect of causing losses to a subset of taxpayers favoured by legislators.
- Second, unanticipated interactions might occur between the legal device and its social environment. That principally means people behaving in ways the designers of the law failed to anticipate, including behaviours resulting from the adoption of new technologies. Autonomous vehicles, for example, will mean that the whole basis for civil liability and insurance in road traffic accidents, the assumption that cars have drivers, might disappear. Instead legal attention will turn to other forms of liability, in particular product liability, but the question will be whether that form of liability is ready for its new role? In the EU, for example, it is not even clear that software counts as a 'product' (Saxby 2016).
- Third, failures might arise from interactions with other legal devices or processes—for example, court decisions might undermine what everyone thought a rule meant, or new statutes might be passed that are not entirely consistent with the previous statute.

As a first cut, one might speculate that problems arising from failures to capture the parties' intentions would come to light early in the life of the statute. In the tax example, it would not be long before the effects become known and complained about. Problems arising from unanticipated interactions with the social environment would occur, one might presume, at a constant rate across the life of the statute. But problems from interactions with other legal processes would tend to rise over time, since, although court cases and new statutes might be random, the number of possible interactions between the statute and cases and new statutes will rise as new cases and statutes accumulate.

On that basis, the typical shape of a legal hazard function would be the familiar bathtub, combining a falling early function, a constant random function and a rising late function. (See Fig. 10.1).

10.3 Sketches for an Empirical Test

The means for easily testing this theory do not yet exist (although work being carried out by the National Archive in London on the UK statutes might at some point give rise to some possibilities).[1] It is difficult to observe faults directly but some proxies might be developed. One possible indication of the existence of a fault is an amendment by a subsequent statute. Admittedly, an amendment is in

[1] http://www.legislation.gov.uk/projects/big-data-for-law.

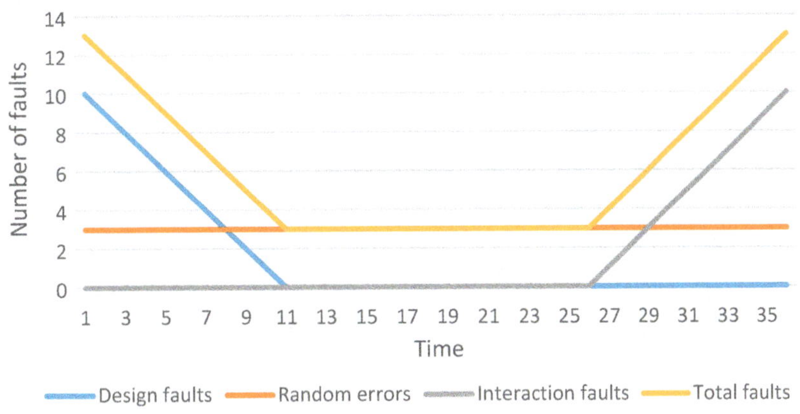

Fig. 10.1 Bathtub curve for law

itself not a fault but rather an attempt at correcting a fault, but if we assume a constant ratio of corrected to uncorrected faults, the amendment rate should be an indication of the underlying fault rate. Another possible indication of a fault is whether a statutory section gives rise to litigation. Because of the cost of litigation, well-designed statutes whose meaning is clear are unlikely to be litigated and so the rate at which people go to court about a statute is an indication of whether something is wrong. Admittedly in the USA, in contrast to the UK, a statute might be litigated not because it is unclear but because it is alleged to be unconstitutional. In that case, the existence of litigation might indicate not a fault in the statute but rather a fault in the constitution. Alternatively, constitutional review might be seen as a 'burn-in' process for statutes.

If we were able to count how many amendments have been made to a statute over time and on how many occasions it has caused litigation, we should be able to construct the statute's empirical hazard curve. At some stage, the National Archive project might succeed in making such counts a fairly straightforward exercise. In the meantime, however, we can do some simple counts, using existing databases for exploratory purposes. For example, we can take a UK statute, the Companies Act 2006, a comprehensive reform and codification of company law, and count the number of amendments made to it by year since it was passed.[2] The result is given in Fig. 10.2.

We can also count the number of cases reported in the law reports that mention the 2006 Act. Only a small proportion of claims come to court and not all cases heard by the courts are officially reported (although the proportion reported in the higher courts is much greater than it was, as a result of it becoming standard practice to store judgments electronically and online). In addition, the mere mention

[2]The count is of changes to the Act and applications of it to new circumstances, but not orders merely bringing parts of the Act into force. UK statutes passed by Parliament having been proposed by ministers usually come into force not when they are passed but by later executive order. Failure to bring a part of an Act into force might be another indication of a design fault, but I have not attempted here to include such faults in the analysis.

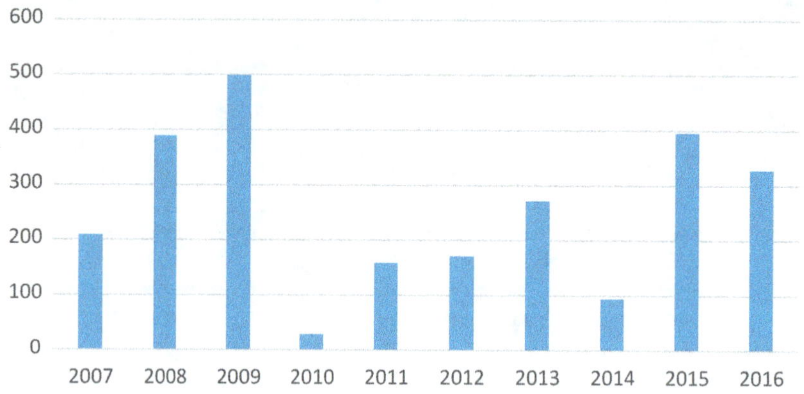

Fig. 10.2 UK Companies Act 2006: number of amendments by year 2007–2016. Source: Author's count using Legislation.co.uk database

of a statute in a case does not necessarily mean that it has caused a problem. But assuming that the proportions of claims to cases coming to court, of cases reported to cases not reported and of problems to mere mentions all remain constant, the number of reported mentions by year is capable of being informative. It should also be remembered that new statutes usually have a long lead-in period with regard to case law. A new statute applies only to situations that arise after it becomes law. Disputes whose facts occurred before the date the new law comes into force, even if the date is after the new law itself came into force, are governed by the law as it stood previously. For that reason, it is useful also to count mentions in the cases of the previous law, in this case the Companies Act 1985. The results are presented in Fig. 10.3.

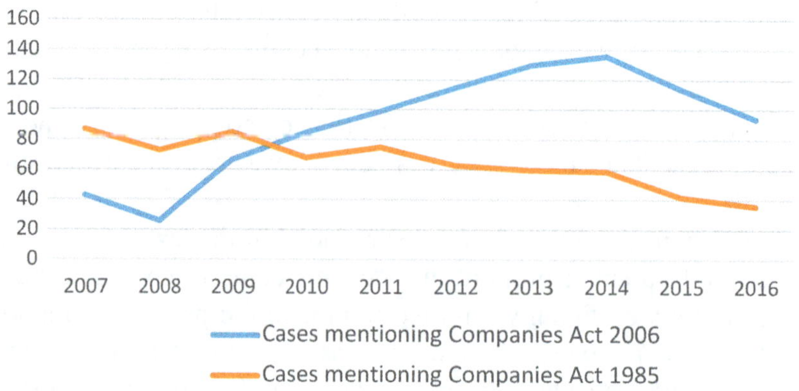

Fig. 10.3 Case law mentions 2007–16 of the Companies Acts 2006 and 1985. Source: Author's count using Westlaw UK database

10.4 Bathtub or Sawtooth or Both?

The pattern for both statutory amendments and case mentions, while not in any way conclusive, seems broadly compatible with the bathtub curve. As the new law comes to be used, problems are identified that result in amendments being needed to correct them. The number of problems then seems to fall off, suggesting a possible move into the 'normal operating' phase, followed by a phase in which interaction with other legal developments begins to cause problems. The case law pattern seems to be a lagged version of the first part of the amendment pattern, a result of the new law only slowly taking over from the old.

The amendment pattern, however, is also compatible with a different kind of cycle, namely the 'sawtooth' pattern proposed for software development (Schick and Wolverton 1978; Pan 1999), in which faults start at a high level and then fall in the test and debugging phase, but rise suddenly with each upgrade before falling to a steady rate as the product becomes obsolete. It is also possible that statutes might follow a combination of the bathtub and the sawtooth, in which faults start high then fall, rise quickly each time the legislature attempts a reform and then fall again, but finally rise as interaction problems come to the fore.

10.5 Possible Future Directions

Much more research is needed before we can even begin to establish the life cycle of statutes. It is possible that statutes about different subjects have different cycles, or that they have similar cycles but of different lengths. It is also possible that different political systems produce different cycles. A system characterised by strong checks and balances such as that of the USA might find statutory amendment difficult, causing faults to find expression more in the case law and less in the legislature, with the whole life cycle taking much longer than in systems, such as that in the UK, characterised by weak checks and balances.

Indeed one final speculation is that constitutions themselves might be subject to life cycles. Thomas Jefferson wanted the US Constitution to be renegotiated every 19 years (Jefferson 1789). Instead, the Constitution has become permanent. But one can perceive in US constitutional history a kind of bathtub. The Bill of Rights, the first ten amendments, are classic design fault corrections. Then random faults come along, including a complete breakdown in the 1860s. After two centuries, however, the Constitution increasingly suffers from interaction problems, especially from incompatible court decisions, incoherently amended statutes and tensions between different levels of government. In other countries, complete replacement of the constitution is not uncommon. France, for example, is on its fifth republic. The USA, however, despite possessing a mechanism for achieving a replacement of its Constitution without breaking legal continuity, in the form of a Convention called under Article 5, chooses to struggle on at the high right-hand side of the curve.

References

Deakin, S., & Koukiadaki, A. (2009). Governance processes, labour management partnership and employee voice in the construction of heathrow terminal 5. *Industrial Law Journal, 38*(4), 365–389.
Gulati, M., & Scott, R. (2013). *The three and a half minute transaction: Boilerplate and the limits of contract design*. Chicago: University of Chicago Press.
Howarth, D. (2014). *Law as engineering: Thinking about what lawyers do*. Cheltenham: Elgar.
Jefferson, T. (1789) Letter to James Madison, 6 September 1789. In J. Boyd et al. (eds.), *The papers of Thomas Jefferson* (Princeton: Princeton University Press), *15*, 392–397.
Nash, F. (2016). *Reliability assessments: Concepts, models, and case studies*. London: Taylor and Francis.
Pan, J. (1999). *Software reliability*. https://users.ece.cmu.edu/~koopman/des_s99/sw_reliability/.
Saxby, S. (2016). *Encyclopedia of information technology law*. London: Sweet and Maxwell.
Schick, G., & Wolverton, R. (1978). An analysis of competing software reliability models. *IEEE Transactions on Software, SE-4*(2).
Simon, H. (1996). *The sciences of the artificial*. Cambridge Mass: MIT Press.
Verma, A., Ajit, S., & Karanki, D. (2016). *Reliability and safety engineering*. London: Springer.

Chapter 11
The Role of Emotion and Culture in the "Moment of Opening"—An Episode of Creative Collaboration

Neeraj Sonalkar and Ade Mabogunje

11.1 Introduction

Creative collaboration in engineering design teams involves interpersonal interactions between the engineers in which the engineers exchange ideas and information. The quality of these interactions determines how ideas are generated, accepted, and evaluated in the team. But what determines the quality of interaction for creative idea generation? What patterns of behavior do we observe during an idea generation interaction? In this chapter, we will look at a particular pattern of interaction called a moment of opening that occurs during idea generation. We will describe the behaviors manifest during a moment of opening and discuss the effect of emotions and culture on a moment of opening interaction.

11.2 Defining Emotion and Culture

In the study of creative collaboration between engineering designers, we are concerned with their in situ behavior. Accordingly, our definition of emotion, and later of culture will entail an attempt to capture the multiple perspectives of the situations and environments engineers typically navigate during design. For the definition of emotion, we rely extensively on the work of Tomkins (1995) and Tsai et al. (2006). Regarding culture, we consider it an amalgamation of biology, biography, geography, and sociology/political-economy. This definition of culture reflects our view

N. Sonalkar · A. Mabogunje (✉)
Center for Design Research, Stanford University, Stanford, CA, USA
e-mail: ade@stanford.edu

N. Sonalkar
e-mail: sonalkar@stanford.edu

E. Subrahmanian et al. (eds.), *Engineering a Better Future*,
https://doi.org/10.1007/978-3-319-91134-2_11

that culture pertains to both an internal environment of an organism and an external environment. Thus, culture involves human–human interaction at both the intrapersonal and the interpersonal levels. It also involves human–external environment interaction both at the level of product engineering and at the level of use, our interest being primarily in the former. What follows is a description of emotion from the two perspectives, which have been of benefit to our work.

11.2.1 Emotion: An Arousal-Valence Perspective

Most studies of emotion in a cultural context tend to refer to data and phenomena signified by emotional words, facial and vocal emotional expressions, and descriptions of emotional experiences. Following Tsai, we subscribe to the notion that these phenomena vary along a minimum of two dimensions: (a) valence and (b) arousal. Tsai gave the example that fear correlates with negative valence and high arousal, and calm correlates with positive valence and low arousal. For Tsai specific emotional states can be called affective states or affect when they are described in terms of valence and arousal (Tsai et al. 2006).

11.2.2 Emotion: An Awareness-Memory Perspective

According to this viewpoint, affect is the innate biological reaction experienced in the body. Feeling is the consciousness of affect. It is a sensation that has been checked against previous experiences and labeled. It is personal and biographical because every person has a distinct set of previous sensations from which to draw when interpreting and labeling his or her feelings. Emotion is the triggering of memories by feelings. It is the projection or display of a feeling. Unlike feelings, the display of emotion can be either genuine or feigned (paraphrased from Nathanson's description of Tomkin's work, 1992).

For Silvan Tomkins, the nine primary affect are:

Positive:

Interest/excitement
Enjoyment/joy

Neutral:

Surprise/startle

Negative:

Distress/anguish
Fear/terror

Anger/rage
Shame/humiliation
Dissmell (smell)
Disgust (taste)

As will be seen later, positive affect is the primary constituent of a moment of opening. Furthermore, in discussing the role of culture, we rely on the idea presented above that; memory or what others call cognitive appraisal will play a role in upregulating or downregulating of emotion (Gross 1998; Ochsner and Gross 2008).

11.3 Moment of Opening as an Episode of Creative Collaboration

The inspiration behind the moment of opening concept comes from creative collaboration in the field of theater. Improvisational theater is a drama form in which the actors do not rehearse predetermined scripts. Instead, they request suggestions from audiences for enacting scenes and collectively improvise a scene on the stage. Improvisational theater has its roots in a series of techniques and exercises created by Keith Johnstone to foster spontaneity and narrative skills among theater actors (Johnstone 1992). The actors use games and collaborative exercises to practice building awareness and listening skills, accepting others' ideas, building on the ideas of others, and being spontaneous with expressing one's own ideas. Since without the script, the actors do not know a priori what they would be doing next, they are open to all possibilities and ideas that are expressed. They accept each other's ideas and add their own ideas to build on them and create a scene together. This interaction results in a group collaborating creatively together. A moment of opening in a design team is an interaction that manifests some of the same patterns visible in an improve group interaction.

A moment of opening can be described as an episode in the interaction between two or more persons characterized by in-the-moment awareness of ideas being exchanged and spontaneous expression of one's ideas. The individuals open up to perceiving ideas from each other and to expressing their own ideas to each other. As observed through video analysis of design teams, this pattern of opening up to each other lasts only for a few minutes at a time. Hence we are calling this pattern of interaction, a moment of opening.

The interaction process in a moment of opening involves three steps.

1. Intake—This involves being aware of what is happening in the team space and listening to the ideas being expressed.
2. Processing—There is free association on the ideas that are listened to by the team member.
3. Expression—The ideas that arise from the free association are spontaneously expressed by the team member.

The following sketch demonstrates the interpersonal interaction that occurs in a moment of opening.

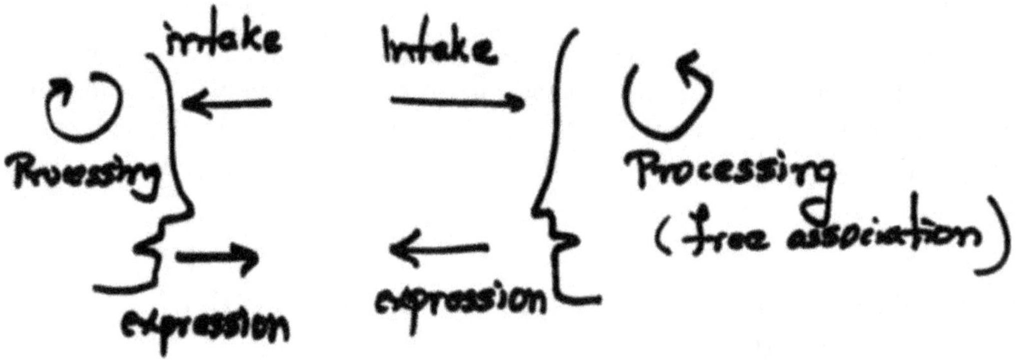

Let us look at the three steps involved in a moment of opening interaction in more detail.

1. Intake—The intake activity involves sensing the environment around the individual to perceive the ideas being expressed. This involves listening, seeing or touching depending on whether the ideas are expressed verbally, through sketching, gesturing, or through artifacts. The intake activity also includes giving validation or positive back-channeling to indicate that the idea has been understood and accepted.
2. Processing—The processing activity involves associating freely on the ideas received. The free association processing is different from other forms of processing like logical analysis, associating on feelings received with the idea or accommodating or assimilating new information, i.e., learning.
3. Expression—The key aspect of expression in a moment of opening interaction is that it is spontaneous. The individual does not filter ideas to be shared with the group based on what is appropriate or not appropriate. In the moment of opening, the response is to express the stream of consciousness that arises from free association in the mind of the individual.

Now, let us look at an example of interaction between three design engineering students that qualifies as a moment of opening. The three students were recruited for a laboratory experiment in which they were given a design task and were instructed to generate ideas for a toy suitable for children 3–7 years old. Their interactions were videotaped and then analyzed for content and emotional cues.

A:	But could it be like at home like when you are a kid you always want to built a tree house or have a tent in the living room you can overnight in	
B:	What about a build your own tent, like a giant like that (gesturing with hand to indicate a big tent)	
A:	Yeah! Yeah! (with excitement)	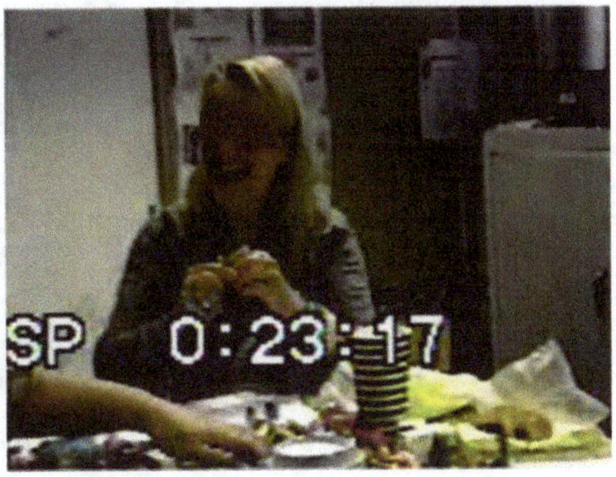

(continued)

(continued)

C:	Build your own tent giant like!	
A:	(repeating A's gesture but exaggerating it to indicate a big tent) so it will like giant tent kit	
C:	Oh that'll be cool!	

In this example, person A starts suggesting an idea for a toy that could capture some of the desires of a child to build a tree house or a tent in the living room. Person B listens to the idea, accepts it, and builds on it to suggest a giant tent. Person B's processing could be considered a free association on the idea as the response did not include judging or expressing feelings or beliefs. Person A listens to B and responds with excitement to the idea. Person C validates the idea being discussed and A repeats the gestures that B used to described the idea and express excitement. In this moment of opening interaction, there is an intake of ideas by A, B, and C, free association on the ideas by B and expression of ideas by A and B.

11.4 The Role of Emotion in a "Moment of Opening" Interaction

We coded the video for emotional cues expressed in the interaction. The following table describes the emotional cues that are visible in the video.

Speaker	Content	Gesture	Tone of voice	Facial expression	Body posture
A:	But could it be like at home like when you are a kid you always want to built a tree house or have a tent in the living room you can overnight in	Hands on the table fiddling with Lego pieces	Neutral ending with an intonation	Neutral	Leaning forward, facing B and turning to C and back to B
B:	What about a build your own tent, like a giant like that (gesturing with hand to indicate a big tent)	Fiddling with Lego pieces and then raising both arms	Neutral	Neutral	Looking from B to C and then to the table
A:	Yeah! Yeah!	Holding hands close to chest, nodding	High arousal	Smiling, open mouth	First leaning forward and then swinging back
C:	Build your own tent giant like!	Fiddling with Lego pieces	Neutral ending with intonation	Smiling	Looking at A
A:	So it will like giant tent kit	Raising arms wide and waving them	High arousal	Smiling, open mouth	Swinging back while gesturing with arms
C:	Oh that'll be cool!	Holding Lego pieces	High arousal	Smiling, open mouth	Looking at A and then to the table

We observe in the video that along with the intake, processing, and expression of the ideas, there is also emotional validation (nodding, saying "yeah", "Oh that'll be cool") and a general synchronicity of emotions among the team members during the interaction. When one team member is excited, the others pick up that emotion and mirror it in their tone, gestures, and facial expression. Thus, a moment of opening interaction can be characterized by certain emotional and ideational patterns. These patterns are described below.

1. Synchronicity of emotions

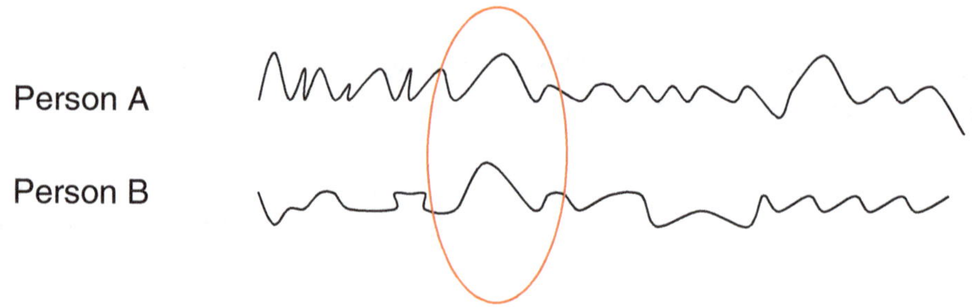

In the example given earlier, we observed that the team members were mirroring each other's emotional expressions and gestures. This mirroring of emotions is called as synchronicity of emotions. If we could obtain the emotional intensity signals of persons A and B, then the emotional synchronicity in the moment of opening would be visible as matching signal patterns. The above figure shows a conceptual visualization of the emotional intensity signals of two persons A and B. The highlighted segment with matching signal patterns indicates a moment of opening.

2. Expressed validation

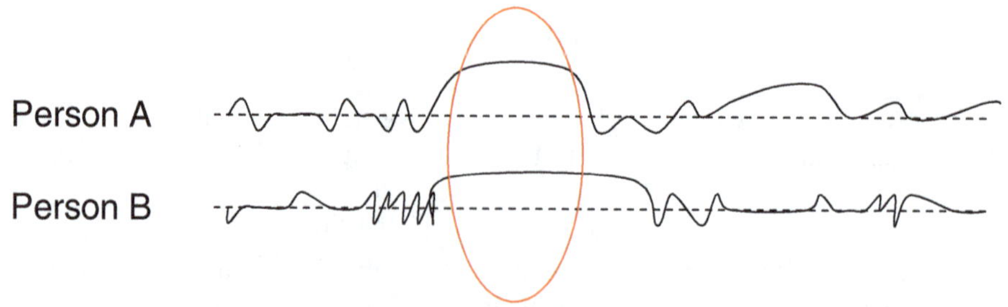

Expressing validation involves giving feedback to the speaker that his idea has been accepted. Validation is expressed through actions like head nodding, paraphrasing the idea or complimenting the idea. In the example discussed earlier, C expresses validation by paraphrasing and complimenting ideas. The above visualization, the highlighted segment shows positive validation from both participants that might indicate a moment of opening.

3. Rapid turn-taking

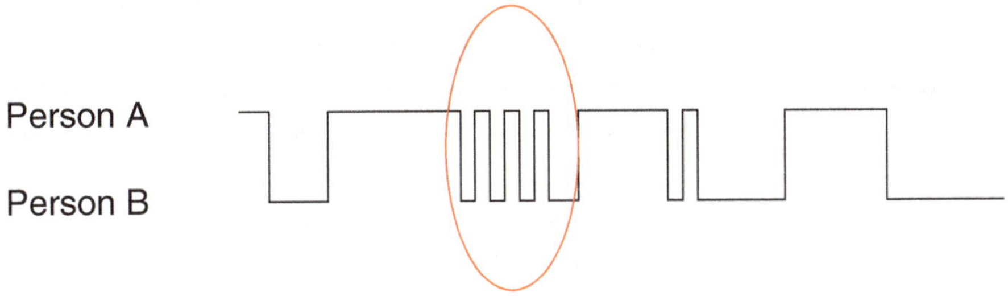

Another characteristic of a moment of opening observed through video analysis is rapid turn-taking. The individuals speak in shorter turns and the speaker turn rotates among the team frequently. The above diagram visualizes the rapid speaker turns occurring between persons A and B as highlighted in a moment of opening.

4. Building on previous ideas

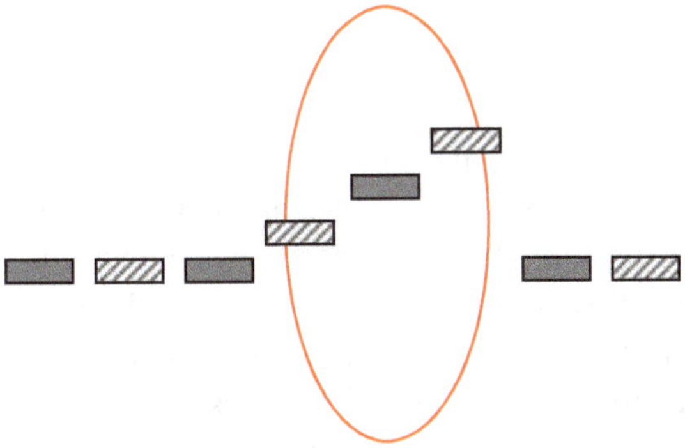

The content of the shorter speaker turns during a moment of opening is not independent. Rather, the individuals build on each other's ideas during the moment of opening. In the conceptual visualization above, the black segments indicate ideas expressed by person A and hatched segments indicate ideas expressed by person B. In the moment of opening highlighted above, the stepped visualization indicates that the ideas and built on each other.

The four patterns of interaction described above are the necessary conditions for a moment of opening to occur during an idea generation session from a team perspective. However, what are the necessary conditions for a moment of opening to occur from an individual perspective?

To answer this question, we utilize the principles of flow behavior as described by Csikszentmihalyi (2001). Csikszentmihalyi described flow as a state of ordered consciousness in which the attention is focused on a task which is challenging enough to match the individual's skills and during which the individual cycles between perceiving external stimulus and acting on the stimulus to achieve preset goals. Csikszentmihalyi further describes flow as a rewarding experience of engaging with the task for the individual. The metaphor of flow could be applied to a moment of opening interaction. The individuals participate in a cycle of perceiving ideas and acting on this stimulus to free associate and express ideas to each other. Csikszentmihalyi described the state of flow occurring when the individual is neither bored by the task because his skills exceed the challenge presented to him, nor anxious because the challenge presented in the task exceeds his skills. Borrowing this concept of bored and anxious states, we can conceptualize that the occurrence of a moment of opening is simultaneous with the individual being neither bored nor anxious during an interaction, but is in a neutral state. This can be visualized as follows.

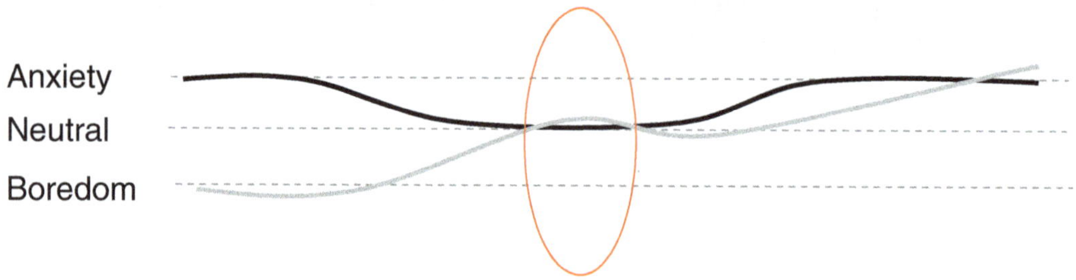

The dark curve indicates the mood of person A during interaction with person B over a period of time. The lighter curve indicates the mood of person B. The individual moods of persons A and B vary between anxiety, neutral, and boredom. It is not yet clear whether being in a neutral state is the precondition for a moment of opening to occur or whether it is a cause of being in a moment of opening. Further research needs to be conducted to investigate this question.

The characteristics of a moment of opening like emotional synchronicity and expressed validation, and the preconditions for individual participation in a moment of opening indicate the importance of the role of emotional expression in a moment of opening interaction. However, what is it that modulates the role of emotional expression in a group interaction? One of the important factors that affect emotional expression is the culture of the individual.

11.5 The Role of Culture in a “Moment of Opening” Interaction

Culture could affect emotional expression in the following ways.

1. Culture influences the display of emotions.
   The emotional cues that an individual displays or perceives from others influence her desire to retreat or advance during an interaction. Positive cues

encourage the individual to continue and negative cues encourage the individual to stop interacting or to change the direction. Culture could determine what emotional cues are displayed, whether positive or negative, because of its influence on the individual's value system. For example, Ekman (1972) found that when Americans and Japanese viewed stressful stimuli first alone and then with a higher status experimenter, the Japanese were more likely to display smiles in the second condition, while there were no differences between Japanese and Americans in emotions displayed in the first condition.

2. Culture determines the range of emotional expression.

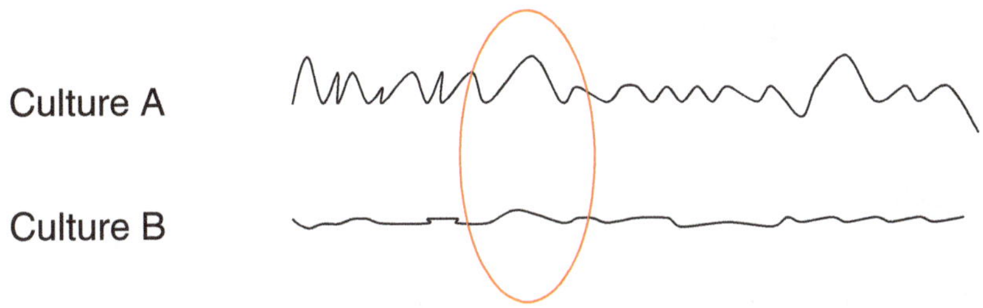

The above diagram shows the emotional intensity signals from persons belonging to two different cultures A and B. Culture A has a wider range of emotional intensity during expression than culture B. So if person A is excited, she would express his excitement by raising her tone of voice, laughing or gesturing wildly. Person B has a smaller range of intensity during emotional expression. If B is equally excited, she might express it by smiling or nodding her head. Thus during a moment of opening, both persons could have emotional synchronicity, but their amplitude could be different. Prior studies about emotional responses in individuals from different cultures support this hypothesis especially for the display for positive affect. For example, Tsai and Levenson (1997) conducted a study about the differences in emotional responses of Chinese Americans and European Americans. The study indicated that Chinese Americans demonstrated less variable and less positive reported affect than European Americans, but did not differ in most measures of physiological responding and in reported negative affect.

3. Culture determines how an emotion is interpreted.

The difference in cultures between two persons interacting during a moment of opening can also affect how they interpret each other's emotions. Prior research conducted by Tsai et al. (2006) at Stanford University on affect valuation in different cultures indicates that cultural factors influence the ideal affect or what people consider as appropriate feeling in different contexts. In the above situation, if person B has a narrower range of emotional intensity in expression, person A might interpret this expression as B being bored or not appropriately excited by the idea, even though B is equally excited about the idea. This could have implication for A's response to B's ideas during the interaction.

11.6 Conclusion

In this chapter, we have described an interaction phenomenon called a moment of opening that occurs during creative collaboration in design teams. We investigated the role of emotion in the moment of opening both at a team level and individual level, and touched briefly on the role of culture in the moment of opening interaction. The concept of moment of opening as an observable interpersonal interaction gave us a practical opening into exploring the influence of emotion and culture in engineering design teams. Further studies into the moment of opening interaction, emotion and culture need to be conducted to broaden our understanding of creative collaboration in design teams.

References

Csikszentmihalyi, M. (2001). *Flow: The psychology of optimal experience*. New York: HarperCollins.

Ekman, P. (1972). Universal and cultural differences in facial expression of emotion. In J. R. Cole (Ed.), *Nebraska symposium on motivation, 1971* (pp. 207–283). Lincoln: Nebraska University Press.

Gross, J. J. (1998). The emerging field of emotion regulation: An integrative review. *Review of General Psychology, 2*, 271–299.

Johnstone, K. (1992). *Impro: Improvisation and the theatre*. New York: Routledge.

Nathanson, D. L. (1992). *Shame and pride: Affect, sex, and the birth of the self*. New York: Norton.

Ochsner, K. N., & Gross, J. J. (2008). Cognitive emotion regulation: Insights from social cognitive and affective neuroscience. *Current Directions in Psychological Science, 17*(2), 153–158.

Tompkins, S. (1995). *Exploring affect: The selected writings of Silvan S. Tompkins*. (V. E. Demos, Ed.). New York: Press Syndicate of the University of Cambridge.

Tsai, J. L., & Levenson, R. W. (1997). Cultural influences on emotional responding: Chinese American and European American dating couples during interpersonal conflict. *Journal of Cross-Cultural Psychology, 28*(5), 600–625.

Tsai, J. L., Knutson, B., & Fung, H. H. (2006). Cultural variation in affect valuation. *Journal of Personality and Social Psychology, 90*(2), 288–307.

Chapter 12
Do the Best Design Ideas (Really) Come from Conceptually Distant Sources of Inspiration?

Joel Chan, Steven P. Dow and Christian D. Schunn

Where do creative design ideas come from? Cognitive scientists have discovered that people inevitably build new ideas from their prior knowledge and experiences (Marsh et al. 1999; Ward 1994). While these prior experiences can serve as sources of inspiration (Eckert and Stacey 1998) and drive sustained creation of ideas that are both new and have high potential for impact (Helms et al. 2009; Hargadon and Sutton 1997), they can also lead designers astray: for instance, designers sometimes incorporate undesirable features from existing solutions (Jansson and Smith 1991; Linsey et al. 2010), and prior knowledge can make it difficult to think of alternative approaches (German and Barrett 2005; Wiley 1998). This raises the question: what features of potential inspirational sources can predict their value (and/or potential harmful effects)? In this chapter, we examine how the conceptual distance of sources relates to their inspirational value.

12.1 Background

12.1.1 Research Base

What do we mean by conceptual distance? Consider the problem of e-waste accumulation: the world generates 20–50 million metric tons of e-waste every year,

J. Chan (✉) · C. D. Schunn
Learning Research and Development Center, University of Pittsburgh, Pittsburgh, PA, USA
e-mail: joc59@pitt.edu

J. Chan
School of Information Studies, University of Maryland, College Park, Maryland, USA

S. P. Dow
Human–Computer-Interaction Institute, Carnegie Mellon University, Pittsburgh, PA, USA

E. Subrahmanian et al. (eds.), *Engineering a Better Future*,
https://doi.org/10.1007/978-3-319-91134-2_12

yielding environmentally hazardous additions to landfills. A designer might approach this problem by building on **near** sources like smaller scale electronics reuse/recycle efforts, or by drawing inspiration from a **far** source like edible food packaging technology (e.g., to design reusable electronics parts). What are the relative benefits of different levels of source conceptual distance along a continuum from near to far?

Many authors, principally those studying the role of analogy in creative problem solving, have proposed that conceptually far sources—structurally similar ideas with many surface (or object) dissimilarities—are the best sources of inspiration for creative breakthroughs (Gentner and Markman 1997; Holyoak and Thagard 1996; Poze 1983; Ward 1998). This proposal—here called the Conceptual Leap Hypothesis—is consistent with many anecdotal accounts of creative breakthroughs, from Kekule's discovery of the structure of benzene by visual analogy to a snake biting its tail (Findlay 1965), to George Mestral's invention of Velcro by analogy to burdock root seeds (Freeman and Golden 1997), to more recent case studies (Enkel and Gassmann 2010; Kalogerakis et al. 2010).

However, empirical support for this proposal is mixed. Some studies have shown an advantage of far over near sources for novelty, quality, and flexibility of ideation (Chan et al. 2011; Chiu and Shu 2012; Dahl and Moreau 2002; Gonçalves et al. 2013; Hender et al. 2002); but, some in vivo studies of creative cognition have not found strong connections between far sources and creative mental leaps (Chan and Schunn 2014; Dunbar 1997), and other experiments have demonstrated equivalent benefits of far and near sources (Enkel and Gassman 2010; Malaga 2000). Relatedly, Tseng et al. (2008) showed that far sources were more impactful after ideation had already begun (vs. before ideation), providing more functionally distinct ideas than near or control, but both far and near sources led to similar levels of novelty. Similarly, Wilson et al. (2010) showed no advantage of far over near sources for novelty of ideas (although near but not far sources decreased variety of ideas). Fu et al. (2013) even found that far sources led to lower novelty and quality of ideas than near sources. Thus, more empirical work is needed to determine whether the Conceptual Leap Hypothesis is well supported. Further, Fu et al. (2013) argue there is an inverted U-shape function in which moderate distance is best, suggesting the importance of conceptualizing and measuring distance along a continuum.

12.1.2 Impetus for the Current Work

Key methodological shortcomings in prior work further motivate more and better empirical work. Prior studies may be too short (typically 30 min to 1 h) to convert far sources into viable concepts. To successfully use far sources, designers must spend considerable cognitive effort to ignore irrelevant surface details, attend to potentially insightful structural similarities, and adapt the source to the target context. Additionally, many far sources may yield shallow or unusable inferences (e.g., due to non-alignable differences in structural or surface features;

Perkins 1997); thus, designers might have to sift through many samples of far sources to find "hidden gems." These higher processing costs for far sources might partially explain why some studies show a negative impact of far sources on the number of ideas generated (Chan et al. 2011; Hender et al. 2002). In the context of a short task, these processing costs might take up valuable time and resources that could be used for other important aspects of ideation (e.g., iteration, idea selection); in contrast, in real-world design contexts, designers typically have days, weeks, or even months (not an hour) to consider and process far sources.

A second issue is a lack of statistical power. Most existing experimental studies have $N \leq 12$ per treatment cell (Chiu and Shu 2012; Hender et al. 2002; Malaga 2000); only four studies had $N \geq 18$ (Chan et al. 2011; Fu et al. 2013; Gonçalves et al. 2013; Tseng et al. 2008), and they are evenly split in support/opposition for the benefits of far sources. Among the few correlational studies, only Dahl and Moreau (2002) had a well powered study design in this regard, with 119 participants and a reasonable range of conceptual distance. Enkel and Gassmann (2010) only examined 25 cases, all of which were cases of cross-industry transfer (thus restricting the range of conceptual distance being considered). This lack of statistical power may have led to a proliferation of false negatives (potentially exacerbated by small or potentially zero effects at short time scales), but possibly also severely overestimated effect sizes or false positives (Button et al. 2013); more adequately powered studies are needed for more precise estimates of the effects of conceptual distance.

A final methodological issue is problem variation. Many experimental studies focused on a single design problem. The inconsistent outcomes in these studies may be partially due to some design problems having unique characteristics, e.g., coincidentally having good solutions that overlap with concepts in far sources. Indeed, Chiu and Shu (2012), who examined multiple design problems, observed inconsistent effects across problems. Other investigations of design stimuli have also observed problem variation for effects (Goldschmidt and Smolkov 2006; Liikkanen and Perttula 2008).

This paper contributes to theories of design inspiration by (1) reporting the results of a study that addresses these methodological issues to yield clearer evidence, and (2) (to foreshadow our results) reexamining theories of design inspiration and conceptual distance in light of accumulating preponderance of evidence *against* the Conceptual Leap Hypothesis.

12.2 Methods

12.2.1 Overview of Research Context

The current work is conducted in the context of OpenIDEO (www.openideo.com), a Web-based crowdsourced innovation platform that addresses a range of social and environmental problems (e.g., managing e-waste, increasing accessibility in elections). The OpenIDEO designers, with expertise in design processes, guide contributors to the platform through a structured design process to produce concepts that are ultimately implemented for real-world impact ("Impact Stories," n.d.). For this study, we focus on three crucial early stages in the process: first, in the *inspiration* phase (lasting between 1.5 and 4 weeks, $M = 3.1$), contributors post *inspirations* (e.g., descriptions of solutions to analogous problems and case studies of stakeholders), which help to define the problem space and identify promising solution approaches; then, in the *concepting* phase (lasting the next 2–6 weeks, $m = 3.4$), contributors post *concepts*, i.e., specific solutions to the problem. Figure 12.1 shows an example concept; it is representative of the typical length and level of detail in concepts, i.e., ~150 words on average, more detail than one or two words/sentences/sketches, but less detail than a full-fledged design report/presentation or patent application. Finally, a subset of these concepts is *shortlisted* by an expert panel (composed of the OpenIDEO designers and a set of domain

E-trash into real cash

Companies can end up with left-over electronics and components for electronics, imagine if there was a marketplace for them to sell their scrap, trash, and left-over chemicals to other companies that need it.

Example Use cases:

1: Big Corp makes 50,000 widgets that need ingredient A in the casing. Unfortunately, the widgets are discontinued and Big Corp is left with mountains of ingredient A that they don't foresee using the future. They are about to throw it all away since they need the space in their warehouse when Big Corp goes to E-trash.com and finds Fancy Corp who just decided to make 100,000 gizmos that really need ingredient A. E-trash facilitates the transaction and mountains of ingredient A don't go to the landfill!

2: Big Corp has thousands of version 1 doodads that they used for the last couple of years but now they need new version 2. They need to get rid of it quickly and so they go to E-trash.com and put it up to find out that Fancy Corp really needs doodads and version 1 works perfectly! Transaction made! Alternatively, version 1 just isn't applicable anymore but ingredient B in it could be very valuable so through E-trash.com they find a recycler who specializes in extracting ingredient B from old electronics and then selling it to other companies.

Description: It would be an on-line marketplace for businesses to find business buyers for their large quantities of e-waste. Sellers could post, either publicly or to select partners, what "waste" they have available and then buyers could bid on the "waste" that they could actually use. Lots of e-products and electronic components can be re-used and re-purposed. This would provide a method for companies to make money off of their waste and to find necessary products and components at a discount. This idea is in large part inspired by the company recyclematch.com. They focus more on traditional manufacturing components.

How does your concept safeguard human health and protect our environment?

It helps to prevent companies from disposing of large quantities of e-waste that other companies could really use.

Where does your concept fit into the lifecycle of electronic devices?

It fits in at the end/beginning of the lifecycle as one business loses the need for the e-waste components or end products that could serve as a foundation point for another company's products.

What steps could be taken today to start implementing your concept?

Encourage existing b2b waste management companies like recyclematch to pursue this by providing the business case of how much money is in working with e-waste especially as the necessary raw materials get harder to find.

What kinds of resources will be needed to fully implement and scale your concept?

Would need an on-line marketplace or perhaps a grant could be awarded to recyclematch or some other player in business to business (b2b) waste management who would already have the necessary connections and framework of a marketplace that could be build upon.

Fig. 12.1 Example concept illustrating the typical amount of detail per concept

experts/stakeholders) for further refinement, based on their creative potential. In later stages, these concepts are refined and evaluated in more detail, and then a subset of them is selected for implementation. We focus on the first three stages given our focus on creative *ideation* (the later stages involve many other design processes, such as prototyping).

The OpenIDEO platform has many desirable properties as a research context for our work, including the existence of multiple design problems, thousands of concepts and inspirations, substantive written descriptions of ideas to enable efficient text-based analyses, and records of feedback received for each idea, another critical factor in design success. A central property for our research question is the explicit nature of sources of inspiration in the OpenIDEO workflow. The site encourages contributors to build on others' ideas. Importantly, when posting concepts or inspirations, contributors are prompted to cite any concepts or inspirations that serve as sources of inspiration for their idea. Also, when browsing other concepts/inspirations, they are also able to see concepts/inspirations the given concept/inspiration "built upon" (i.e., cited as explicit sources of inspiration; see Fig. 12.2). This culture of citing sources is particularly advantageous, given that people generally forget to monitor or cite their sources of inspiration (Brown and Murphy 1989; Marsh et al. 1997), and our goal is to study the effects of source use. While users might still forget to cite sources, these platform features help ensure higher rates of source monitoring than other naturalistic ideation contexts. We note that this operationalization of sources as self-identified citations precludes consideration of implicit stimulation; however, the Conceptual Leap Hypothesis may be more applicable to conscious inspiration processes (e.g., analogy, for which conscious processing is arguably an important defining feature; Schunn and Dunbar 1996).

Fig. 12.2 Depiction of OpenIDEO citation workflow. When posting concepts/inspirations, users are prompted to cite concepts/inspirations they "build upon" by dragging bookmarked concepts/inspirations (middle panel) to the citation area (left panel). Users can also search for related concepts/inspirations at this step (middle panel). These cited sources then show up as metadata for the concept/inspiration (right panel)

12.2.2 Sample and Initial Data Collection

The full dataset for this study consists of 2341 concepts posted for 12 completed challenges by 1190 unique contributors, citing 4557 unique inspirations; 241 (10%) of these concepts are shortlisted for further refinement. See Table 12.2 for a description of the 12 challenges (with some basic metadata on each challenge). Figure 12.3 shows the full-text design brief for two challenges.

With administrator permission, we downloaded all inspirations and concepts (which exist as individual webpages) and used an HTML parser to extract the following data and metadata:

(1) Concept/inspiration author (who posted the concept/inspiration)
(2) Number of comments (before the refinement phase)
(3) Shortlist status (yes/no)
(4) List of cited sources of inspiration
(5) Full-text of concept/inspiration.

How can we manage e-waste & discarded electronics to safeguard human health & protect our environment?

Ever wondered what happens to your outmoded cell phone when you replace it with the latest model? Or where a battery goes when you toss it in the trash? Around the world, end-of-life electronics that are waste, also known as e-waste, present a significant challenge for our environment and our health. Together with Brazilian bank Itaú Unibanco, the U.S. Department of State, and the U.S. Environmental Protection Agency, we're asking the OpenIDEO community to help us find ways to manage e-waste to better safeguard human health and protect our environment.According to the Consumer Electronics Association, in 2012 global spending on electronics is expected to surpass US$1 trillion. As use of these electronics increases around the world, the question of how to properly manage them when consumers are finished with them becomes more urgent. Unfortunately, not enough of our electronics are re-used, recycled or refurbished. Too many of them end up directly in landfills or recovered in an unsafe manner. According to the UN Environmental Programme, some 20 to 50 million metric tons of e-waste are generated worldwide every year, with mobile phones and televisions contributing 10 million tons per year by 20152. E-waste presents complex issues with many factors to consider, one of them being the environmental impact of hazardous substances and toxic chemicals – including lead, nickel, cadmium and mercury. The great news is that many of the materials used in consumer electronics can be recycled or refurbished to be used in other electronics. The opportunity is to find better ways to manage our used and end-of-life electronics and avoid them ending up in landfills.

How might we increase the number of registered bone marrow donors to help save more lives?

OpenIDEO has partnered with the Haas Center for Public Service at Stanford University to explore new ideas for encouraging bone marrow donation worldwide. Together we're asking you, the OpenIDEO community, to help us find ways to expand the global network of potential bone marrow donors and support people who are battling leukemia and other blood cancers. Bone marrow transplants are one type of treatment for leukemia and other blood or bone marrow cancers. This OpenIDEO challenge will complement the efforts of 100K Cheeks, a Stanford-based advocacy group dedicated to increasing the number of people enrolled in bone marrow registries worldwide. Certain populations are dramatically under-represented in existing bone marrow registries. For example, the match rate within the South Asian* demographic is critically low—with a 1 in 20,000 chance for a potential recipient to find a match. For more information about bone marrow donation (including the process, myths, and facts), visit BeTheMatch.org. If you're interested in becoming a donor, you can look up the registry in your country here.

Fig. 12.3 Full-text of challenge briefs from two OpenIDEO challenges

Not all concepts cited inspirations as sources. Of the 2341 concepts, 707 (posted by 357 authors) cited at least one inspiration, collectively citing 2245 unique inspirations. 110 of these concepts ($\sim 16\%$) were shortlisted (see Table 12.1 for a breakdown by challenge). This set of 707 concepts is the primary sample for this study; the others serve as a contrast to examine the value of explicit building at all on prior sources, and to aid in interpretation of any negative or positive effects of variations in distance. Because we only collected publicly available data, we do not have complete information on the expertise of all contributors: however, based on their public profiles on OpenIDEO, at least 1/3 of the authors in this sample are professionals in design-related disciplines (e.g., user experience/interaction design, communication design, architecture, product/industrial design, entrepreneurs and social innovators, etc.) and/or domain experts or stakeholders (e.g., urban development researcher contributing to the vibrant-cities challenge, education policy researcher contributing to the youth-employment challenge, medical professional contributing to the bone marrow challenge). Collectively, these authors accounted for approximately half of the 707 concepts in this study.

Table 12.1 Descriptions and number of posts for OpenIDEO challenges in final analysis sample

Name/description	# of inspirations	# of concepts (shortlisted)
How might we increase the number of registered bone marrow donors to help save more lives?	186	71 (7)
How might we inspire and enable communities to take more initiative in making their local environments better?	160	44 (11)
How can we manage e-waste and discarded electronics to safeguard human health and protect our environment?	60	26 (8)
How might we better connect food production and consumption?	266	147 (10)
How can technology help people working to uphold human rights in the face of unlawful detention?	248	62 (7)
How might we identify and celebrate businesses that innovate for world benefit and inspire other companies to do the same?	122	24 (13)
How might we use social business to improve health in low-income communities?	131	46 (11)
How might we increase social impact with OpenIDEO over the next year?	67	40 (12)
How might we restore vibrancy in cities and regions facing economic decline?	558	119 (13)
How might we design an accessible election experience for everyone?	241	47 (8)
How might we support web entrepreneurs in launching and growing sustainable global businesses?	88	49 (7)
How can we equip young people with the skills, information and opportunities to succeed in the world of work?	118	32 (3)

We analyze the impact of the distance of inspirations (and not cited concepts) given our focus on ideation processes during "original" or nonroutine design, where designers often start with a problem and only "inspirations" (e.g., information about the problem or potentially related designs) rather than routine design (e.g., configuration or parametric design), where designers might be modifying or iterating on existing solutions rather than generating novel ones (Chakrabarti 2006; Dym 1994; Gero 2000; Ullman 2002). The Conceptual Leap Hypothesis maps most clearly to nonroutine design.

12.2.3 Measures

12.2.3.1 Creativity of Concepts

We operationalize concept creativity as whether a concept gets shortlisted. Shortlisting is done by a panel of expert judges, including the original challenge sponsors, who have spent significant time searching for and learning about existing approaches, and the OpenIDEO designers, who are experts in the general domain of creative design, and who have spent considerable time upfront with challenge sponsors learning about and defining the problem space for each challenge.

An expert panel is widely considered a "gold standard" for measuring the creativity of ideas (Amabile 1982; Baer and McKool 2009; Brown 1989; Sawyer 2012). Further, we know from conversations with the OpenIDEO team that the panel's judgments combine consideration of both novelty and usefulness/appropriateness (here operationalized as potential for impact; A. Jablow, personal communication, May 1, 2014), the standard definition of creativity (Sawyer 2012). Since OpenIDEO challenges are novel and unsolved, successful concepts are different from (and, perhaps more importantly, significantly better than) the existing unsatisfactory solutions. We use shortlist (rather than win status) given our focus on the ideation phase in design (vs. convergence/refinement, which happens after concepts are shortlisted, and can strongly influence which shortlisted concepts get selected as "winners" for implementation).

12.2.3.2 Conceptual Distance

Measurement Approach

Measuring conceptual distance is a major methodological challenge, especially when studying large samples of ideation processes (e.g., many designs across many design problems). The complex and multifaceted nature of typical design problems can make it difficult to distinguish "within" and "between" domain sources in a consistent and principled manner. Further, using only a binary scale risks losing variance information that could be critical for converging on a more precise understanding of the effects of conceptual distance (e.g., curvilinear effects across the continuum of distance). Continuous distance measures are an attractive alternative, but can be extremely costly to obtain at this scale, especially for naturalistic sources (e.g., relatively developed text descriptions vs. simple sketches or one-to-two sentence descriptions). Human raters may suffer from high levels of fatigue, resulting in poor reliability or drift of standards.

We address this methodological challenge with probabilistic topic modeling (Blei 2012; Steyvers and Griffiths 2007), a major computational approach for understanding large collections of unstructured text. They are similar to other unsupervised machine learning methods—e.g., K-means clustering, and Latent Semantic Analysis (Deerwester et al. 1990)—but distinct in that they emphasize human understanding of not just the relationship between documents in a collection, but the "reasons" for the hypothesized relationships (e.g., the "meaning" of particular dimensions of variation), largely because the algorithms underlying these models tend to produce dimensions in terms of clusters of tightly co-occurring words. Thus, they have been used most prominently in applications where understanding of a corpus, not just information retrieval performance, is a high priority goal, e.g., knowledge discovery and information retrieval in repositories of scientific papers (Griffiths and Steyvers 2004), describing the structure and evolution of scientific fields (Blei and Lafferty 2006, 2007), and discovering topical dynamics in social media use (Schwartz et al. 2013).

We use Latent Dirichlet Allocation (LDA; Blei et al. 2003), the simplest topic model. LDA assumes that documents are composed of a mixture of latent "topics" (occurring with different "weights" in the mixture), which in turn generate the words in the documents. LDA defines topics as probability distributions over words: for example, a "genetics" topic can be thought of as a probability distribution over the words {phenotype, population, transcription, cameras, quarterbacks}, such that words closely related to the topic {phenotype, population, transcription} have a high probability in that topic, and words not closely related to the topic {cameras, quarterbacks} have a very low probability. Using Bayesian statistical learning algorithms, LDA infers the latent topical structure of the corpus from the co-occurrence patterns of words across documents. This topical structure includes 1) the topics in the corpus, i.e., the sets of probability distributions over words, and 2) the topic mixtures for each document, i.e., a vector of weights for each of the corpus topics for that document. We can derive conceptual *similarity* between any pair of documents by computing the cosine between their topic-weight vectors. In essence, documents that share dominant topics in similar relative proportions are the most similar.

Here, we used the open-source MAchine Learning for LanguagE Toolkit (MALLET; McCallum 2002) to train an LDA model with 400 topics for all documents in the full dataset, i.e., 2341 concepts, 4557 inspirations, and 12 challenge briefs (6910 total documents). Additional technical details on the model-building procedure are available in Appendix 1. Resulting cosines between inspirations and the challenge brief ranged from 0.01 to 0.91 ($M = 0.21$, $SD = 0.18$), a fairly typical range for large-scale information retrieval applications (Jessup and Martin 2001).

Validation

Since we use LDA's measures of conceptual distance as a *substitute* for human judgments, we validate the adequacy of our topic model using measures of fit with human similarity judgments on a subset of the data by trained human raters.

Five trained raters used a Likert-type scale to rate 199 inspirations from one OpenIDEO challenge for similarity to their challenge brief, from 1 (very dissimilar) to 6 (extremely similar). Raters were given the intuition that the rating would approximately track the proportion of "topical overlap" between each inspiration and the challenge brief, or the extent to which they are "about the same thing." The design challenge context was explicitly deemphasized, so as to reduce the influence of individual differences on perceptions of the "relevance" of sources of inspiration. Thus, the raters were instructed to treat all the documents as "documents" (e.g., an article about some topics, vs. "problem solution") and consciously avoid judging the "value" of the inspirations, simply focusing on semantic similarity. Raters listed major topics in the challenge brief and evaluated each inspiration against those major topics. To ensure internal consistency, the raters also sorted the inspirations by similarity after every 15–20 judgments. They then inspected the rank ordering and composition of inspirations at each point in the scale, and made adjustments if necessary (e.g., if an inspiration previously rated as "1" now, in light of newly encountered inspirations, seemed more like a "2" or "3"). Although the task was difficult, the mean ratings across raters had an acceptable aggregate consistency intra-class correlation coefficient [ICC(2,5)] of 0.74 (mean inter-coder correlation = 0.36). LDA cosines correlated highly, at $r = 0.51$, 95% CI = [0.40, 0.60], with the continuous human similarity judgments (see Fig. 12.4A). We note that this correlation is better than the highest correlation between human raters ($r = 0.48$), reinforcing the value of automatic coding methods for this difficult task.

For comparability with prior work, we also measure fit with binary (within- vs. between-domain) distance ratings. Two raters also classified 345 inspirations from a different challenge as either within- or between-domain. Raters first collaboratively defined the problem domain, focusing on the question, "What is the problem to be solved?" before rating inspirations. Within-domain inspirations were information about the problem (e.g., stakeholders, constraints) and existing prior solutions for very similar problems, while between-domain inspirations were information/solutions for analogous or different problems. Reliability for this measure was acceptable, with an overall average kappa of 0.78 (89% agreement). All disagreements were resolved by discussion. Similar to the continuous similarity judgments, the point biserial correlation between the LDA-derived cosine and the binary judgments was also high, at 0.50, 95% CI = [0.42, 0.58]. The mean cosine to the challenge brief was also higher for within-domain ($M = 0.49$, SD = 0.25, $N = 181$) vs. between-domain inspirations ($M = 0.23$, SD = 0.20, $N = 164$), $d = 1.16$, 95% CI = [1.13, 1.19] (see Fig. 12.4b), further validating the LDA approach to measuring distance. Figure 12.5 shows examples of a near and far inspiration (from the e-waste challenge), along with the top 3 LDA topics (represented by the top 5 words for that latent topic), computed cosine vs. its challenge

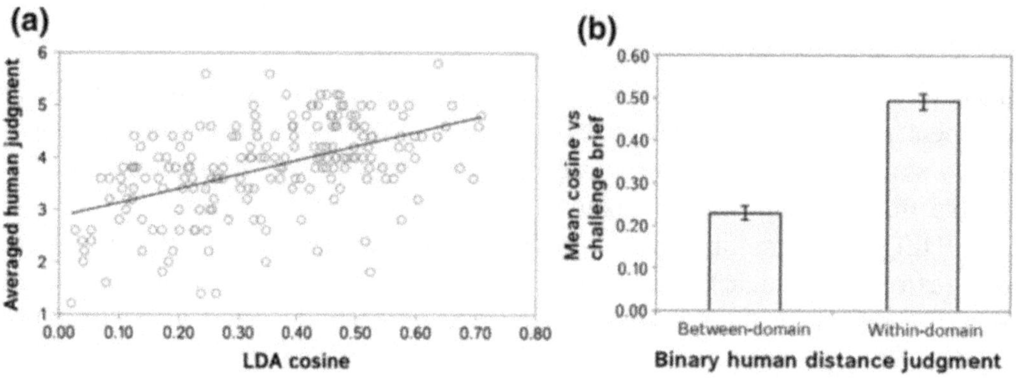

Fig. 12.4 **a** Scatterplot of LDA cosines vs. averaged human continuous similarity judgments for inspirations in the e-waste challenge. **b** Mean cosine against the challenge brief for within—versus between-domain inspirations

brief, and human similarity rating. The top 3 topics for the challenge brief are {waste, e, recycling, electronics, electronic}, {waste, materials, recycling, recycled, material}, and {devices, electronics, electronic, device, products}, distinguishing e-waste, general recycling, and electronics products topics. These examples illustrate how LDA is able to effectively extract the latent topical mixture of the inspirations from their text (inspirations with media also include textual descriptions of the media, mitigating concerns about loss of semantic information due to using only text as input to LDA) and also capture intuitions about variations in conceptual distance among inspirations: a document about different ways of assigning value to possessions is intuitively conceptually more distant from the domain of e-waste than a document about a prior effort to address e-waste.

The near and far examples depicted in Fig. 12.5 also represent the range of conceptual distance measured in this dataset, with the near inspiration's cosine of 0.64 representing approximately the 90th percentile of similarity to the challenge domain, and the far inspiration's cosine of 0.01 representing approximately the 10th percentile of similarity to the challenge domain. Thus, the range of conceptual distance of inspirations in this data spans approximately from sources that are very clearly within the domain (e.g., an actual solution for the problem of electronic waste involving recycling of materials) to sources that are quite distant, but not obviously random (e.g., an observation of how people assign emotional value to relationships and artifacts). This range most likely excludes the "too far" example designs studied in Fu et al. (2013) or the "opposite stimuli" used in Chiu and Shu (2012).

Final Distance Measures

The challenge briefs varied in length and specificity across challenges, as did mean raw cosines for inspirations. But, these differences in mean similarity were much larger, d = 1.90, 95% CI = [1.85–1.92] (for 80 inspirations from 4 challenges with maximally different mean cosines), than for human similarity judgments (coded

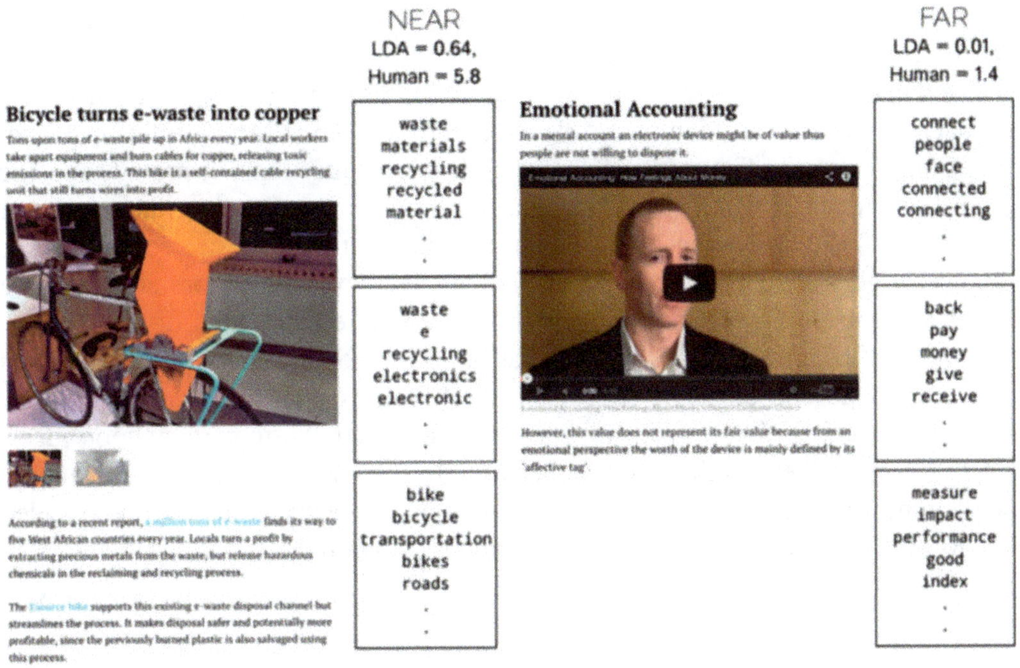

Fig. 12.5 Topics found by LDA within examples of near and far inspirations for the e-waste challenge

separately but with the same methodology as before), $d = 0.18$, 95% CI = [–0.05 to 0.43]. This suggested that between-challenge differences were more an artifact of variance in challenge brief length/specificity. Thus, to ensure meaningful comparability across challenges, we normalized the cosines by computing the z-score for each inspiration's cosine relative to other inspirations from the same challenge before analyzing the results in the full dataset. However, similar results are found using raw cosines, but with more uncertainty in the statistical coefficient estimates.

We then subtracted the cosine z-score from zero such that larger values meant more distant. From these "reversed" cosine z-scores, two different distance measures were computed to tease apart possibly distinct effects of source distance: (1) *max* distance ($DIST_{MAX}$), i.e., the distance of a concept's furthest source from the problem domain and (2) *mean* distance ($DIST_{MEAN}$) of the concept's sources. $DIST_{MAX}$ estimates "upper bounds" for the benefits of distance: do the best ideas really come from the furthest sources? $DIST_{MEAN}$ capitalizes on the fact that many concepts relied on multiple inspirations and estimates the impact of the relative *balance* of relying on near vs. far sources (e.g., more near than far sources, or vice versa).

12.2.3.3 Control Measures

Given our correlational approach, it is important to identify and rule out or adjust for other important factors that may influence the creativity of concepts (particularly in the later stages, where prototyping and feedback are especially important) and may be correlated with the predictor variables.

Feedback. Given the collaborative nature of OpenIDEO, we reasoned that feedback in the form of comments (labeled here as *FEEDBACK*) influences success. Comments can offer encouragement, raise issues/questions, or provide specific suggestions for improvement, all potentially significantly enhancing the quality of the concept. Further, feedback may be an alternate pathway to success via source distance, in that concepts that build on far sources may attract more attention and therefore higher levels of feedback, which then improve the quality of the concept.

Quality of cited sources. Concepts that build on existing high-quality concepts (e.g., those who end up being shortlisted or chosen as winners) have a particular advantage of being able to learn from the mistakes and shortcomings, good ideas, and feedback in these high-quality concepts. Thus, as a proxy measure of quality, the number of shortlisted concepts a given concept builds upon (labeled *SOURCESHORT*) could be a large determinant of a concept's success.

12.2.4 Analytic Approach

We are interested in predicting the creative outcomes of 707 concepts, posted by 357 authors for 12 different design challenges. Authors are not cleanly nested within challenges, nor vice versa; our data are cross-classified, with concepts cross-classified within both authors and challenges (see Fig. 12.6). This cross-classified structure violates assumptions of uniform independence between concepts: concepts posted by the same author or within the same challenge may be more similar to each other. Failing to account for this nonindependence could lead to overestimates of the statistical significance of model estimates (i.e., make unwarranted claims of statistically significant effects). This issue is exacerbated

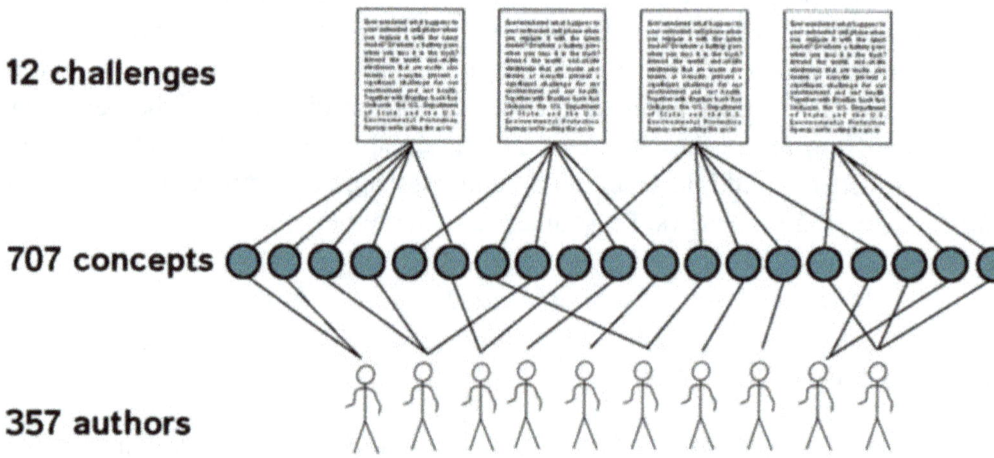

Fig. 12.6 Illustrated cross-classified structure of the data

when testing for small effects. Additionally, modeling between-author effects allows us to separate author effects (e.g., higher/lower creativity) from the impact of sources on individual concepts. Thus, we employ generalized linear mixed models (also called hierarchical generalized linear models) to model both fixed effects (of our independent and control variables) and random effects (potential variation of the outcome variable attributable to author- or challenge-nesting and also potential between-challenge variation in the effect of distance) on shortlist status (a binary variable, which requires logistic, rather than linear, regression).

An initial model predicting the outcome with only the intercept and between-challenge and -author variation confirms the presence of significant non-independence, with between-author and between-challenge variation in shortlist outcomes estimated at 0.44, and 0.50, respectively. The intra-class correlations for author-level and challenge-level variance in the intercept are $\sim$0.11 and 0.13, respectively, well above the cutoff recommended by Raudenbush and Bryk (2002).[1]

12.3 Results

12.3.1 Descriptive Statistics

On average, 16% of concepts in the sample get shortlisted (see Table 12.2). $DIST_{MEAN}$ is centered approximately at 0, reflecting our normalization procedure. Both $DIST_{MAX}$ and $DIST_{MEAN}$ have a fair degree of negative skew. *SOURCESHORT* and *FEEDBACK* have strong positive skew (most concepts either have few comments or cite 0 or 1 shortlisted concepts).

There is a strong positive relationship between $DIST_{MAX}$ and $DIST_{MEAN}$ (see Table 12.3). All variables have significant bivariate correlations with *SHORTLIST* except for $DIST_{MAX}$; however, since it is a substantive variable of interest, we will model it nonetheless. Controlling for other variables might enable us to detect subtle effects.

12.3.2 Statistical Models

We estimated separate models for the effects of $DIST_{MAX}$ and $DIST_{MEAN}$, each controlling for challenge-and author-nesting, *FEEDBACK*, and *SHORTSOURCE*.

[1]Although concept-level variance is not estimated in mixed logistic regressions, we follow Zeger et al. (1988) suggestion of $(15/16)\pi^3/3$ as a reasonable approximation for residual level-1 variance (the concept level in our case).

Table 12.2 Descriptive statistics

Variable	Valid N	Min	Max	Mean	Median	SD
SHORTLIST	707	0.00	1.00	0.16	0.00	0.36
$DIST_{MAX}$	707	−3.85	1.90	0.45	0.76	0.85
$DIST_{MEAN}$	707	−3.85	1.67	−0.10	0.01	0.85
SOURCESHORT	707	0	11	0.51	0	0.96
FEEDBACK	707	0	67	8.43	6	9.45

Table 12.3 Bivariate correlations

Variable	$DIST_{MAX}$	$DIST_{MEAN}$	*SOURCESHORT*	*FEEDBACK*
SHORTLIST	−0.05	−0.10*	0.11**	0.33***
$DIST_{MAX}$		0.77***	0.05	0.07 ^{m}
$DIST_{MEAN}$			−0.05	0.01
SOURCESHORT				0.12**

$^{m}p < .10$, $*p < .05$, $**p < .01$, $***p < .001$

12.3.2.1 Max Distance

Our model estimated an inverse relationship between $DIST_{MAX}$ and Pr (shortlist), such that a 1-unit increase in $DIST_{MAX}$ predicted a 0.33 *decrease* in the log-odds of being shortlisted, after accounting for the effects of *FEEDBACK, SHORTSOURCE,* and challenge- and author-level nesting, $p < .05$ (see Appendix 2 for technical details on the statistical models). However, this coefficient was estimated with considerable uncertainty, as indicated by the large confidence intervals (coefficient could be as small as −0.06 or as large as −0.60); considering also the small bivariate correlation with *SHORTLIST*, we are fairly certain that the "true" coefficient is *not* positive (*contra* the Conceptual Leap Hypothesis), but we are quite uncertain about its magnitude.

Figure 12.7 visually displays the estimated relationship between $DIST_{MAX}$ and Pr (shortlist), evaluated at mean values of feedback and shortlisted sources. To aid interpretation, we also plot the predicted Pr (shortlist) for concepts that cite no sources using a horizontal gray bar (bar width indicates uncertainty in estimate of Pr (shortlist)): concepts with approximately equivalent amounts of feedback (i.e., mean of 8.43) have a predicted Pr (shortlist = 0.09, 95% CI = [0.07 to 0.11]; using a logistic model, the coefficient for "any citation" (controlling for feedback) is 0.31, 95% CI = [0.01 to 0.62]). This bar serves as an approximate "control" group, allowing us to interpret the effect not just in terms of the effects of far sources relative to near sources, but also in comparison with using no sources. Comparing the fitted curve with this bar highlights how the advantage of citing versus not citing inspirations seems to be driven mostly by citing relatively near inspirations: Pr (shortlist) for concepts that cite far inspirations converges on that of no-citation

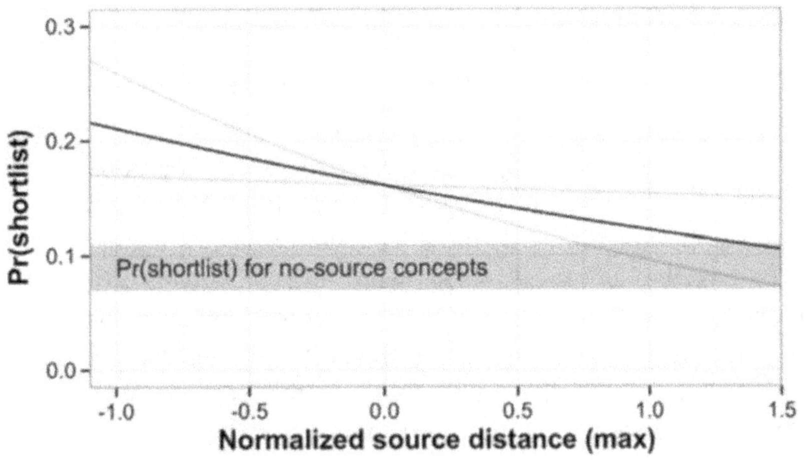

Fig. 12.7 Model-fitted relationship between $DIST_{MAX}$ and Pr (shortlist), evaluated at mean values of feedback and source shortlist. Grayed lines are fits with upper and lower limits for 95% CI for effect of $DIST_{MAX}$

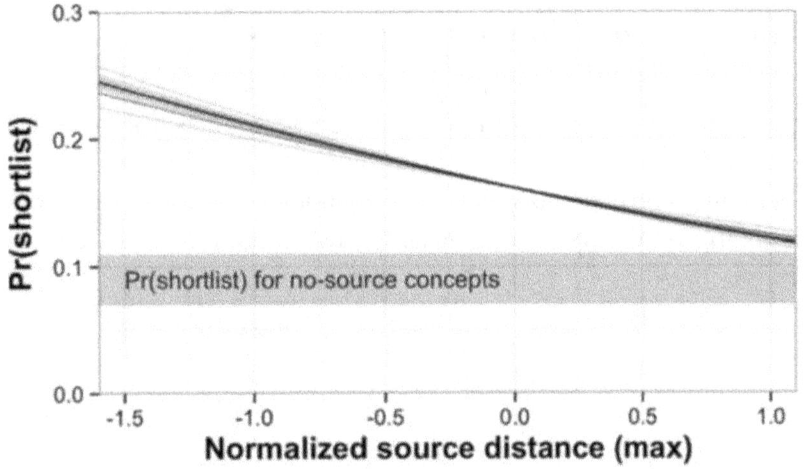

Fig. 12.8 Overall and by-challenge model-fitted relationship between $DIST_{MAX}$ and Pr (shortlist). Fitted values evaluated at mean values of feedback and source shortlist. Grayed lines are fits for each individual challenge

concepts. We emphasize again that despite the uncertainty in the *degree* of the negative relationship between $DIST_{MAX}$ and Pr (shortlist), the data do *not* support an inference that the best ideas are coming from the farthest inspirations: rather, relying on nearer rather than farther sources seems to lead to more creative design ideas. Importantly, this pattern of results was robust across challenges on the platform: the model estimated essentially zero between-challenge variation in the slope of $DIST_{MAX}$. $\chi^2(2) = 0.05$, $p = 0.49$ (see Fig. 12.8).

12.3.2.2 Mean Distance

Similar results were obtained for $DIST_{MEAN}$. There was a robust inverse relationship between $DIST_{MEAN}$ and Pr (shortlist), such that a 1-unit increase in $DIST_{MEAN}$ was associated with a *decrease* of approximately 0.40 in the log-odds of being shortlisted, $p < .05$. The estimates of this effect were obtained with similarly low precision regarding the magnitude of the effect, with 95% CI upper limit of at most $B = -0.09$ (but as high as -0.71). As shown in Fig. 12.9, as $DIST_{MEAN}$ increases, Pr (shortlist) approaches that of non-citing concepts, again suggesting (as with $DIST_{MAX}$) that the most beneficial sources appear to be ones that are relatively close to the challenge domain. Again, as with $DIST_{MAX}$, this pattern of results did not vary across challenges: our model estimated essentially zero between-challenge variation in the slope of $DIST_{MEAN}$, $\chi^2(2) = 0.07$, $p = .48$ (see Fig. 12.10).

12.4 Discussion

12.4.1 Summary and Interpretation of Findings

This study explored how the inspirational value of sources varies with their conceptual distance from the problem domain along the continuum from near to far. The study's findings provide no support for the notion that the best ideas come from building explicitly on the farthest sources. On the contrary, the benefits of building

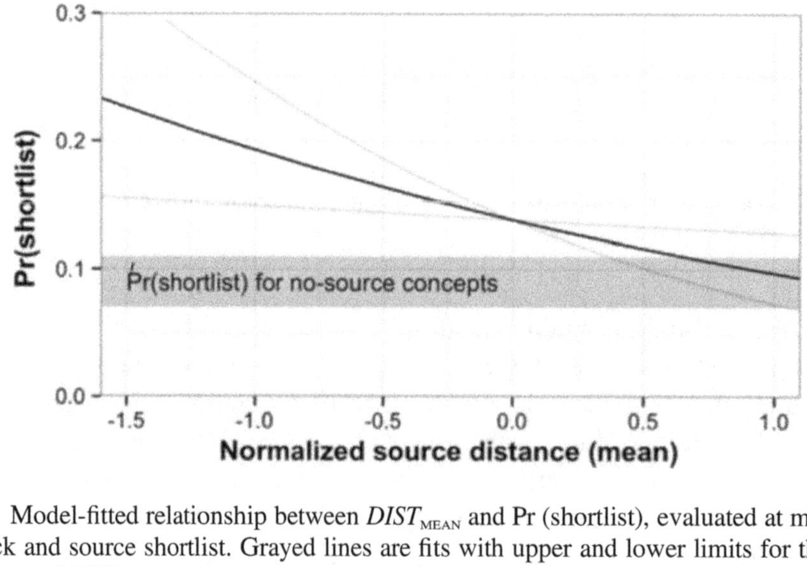

Fig. 12.9 Model-fitted relationship between $DIST_{MEAN}$ and Pr (shortlist), evaluated at mean values of feedback and source shortlist. Grayed lines are fits with upper and lower limits for the 95% CI for the effect of $DIST_{MEAN}$

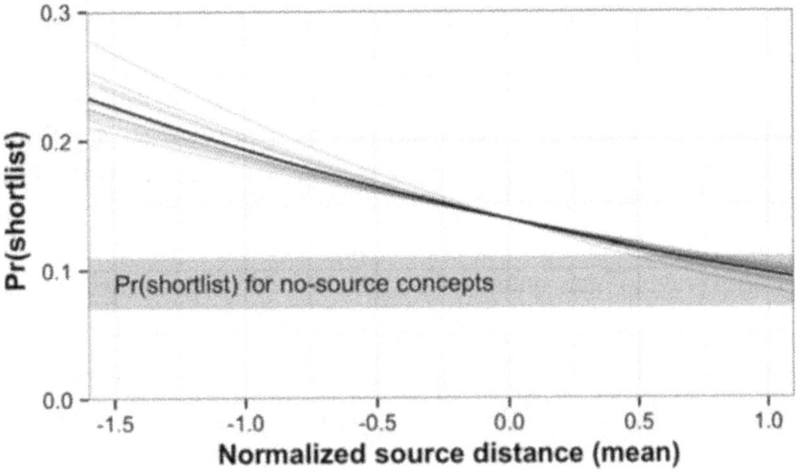

Fig. 12.10 Overall and by-challenge model-fitted relationship between $DIST_{MEAN}$ and Pr (shortlist). Fitted values evaluated at mean values of feedback and source shortlist. Grayed lines are fits for each individual challenge

explicitly on inspirations seem to accrue mainly for concepts that build more on near than far inspirations. Importantly, these effects were consistently found in all of the challenges, addressing concerns raised about potential problem variation, at least among nonroutine social innovation design problems.

12.4.2 *Caveats and Limitations*

Some caveats should be discussed before addressing the implications of this study. First, the statistical patterns observed here are conditional: i.e., we find an inverse relationship between conceptual distance of *explicitly cited* inspiration sources and Pr (shortlist). Our data are silent on the effects of distance for concepts that did not cite sources (where lack of citation could indicate forgetting of sources or lack of conscious building on sources).

There is a potential concern over range restriction or attrition due to our reliance on self-identified sources. However, several features of the data help to ameliorate this concern. First, concepts that did not cite sources were overall of lower quality; thus, it is unlikely that the inverse effects of distance are solely due to attrition (e.g., beneficial far inspirations not being observed). Second, the integration of citations and building on sources into the overall OpenIDEO workflow and philosophy of ideation also helps ameliorate concerns about attrition of far sources. Finally, the dataset included many sources that were quite far away, providing sufficient data to statistically test the effects of relative reliance on far sources (even if they are overall underreported). Nevertheless, we should still be cautious about making inferences about the impact of *unconscious* sources (since sources in this data are explicitly cited and therefore consciously built upon). However, as we note in the methods, the Conceptual Leap Hypothesis maps most cleanly to conscious inspiration processes (e.g., analogy).

Finally, some may be concerned that we have not measured novelty here. Conceivably, the benefits of distance may only be best observed for the novelty of ideas, and not necessarily quality, consistent with some recent work (Franke et al. 2013). However, novelty *per se* does not produce creativity; we contend that to fully understand the effects of distance on design creativity, we must consider its impacts on both novelty and quality together (as our shortlist measure does).

12.4.3 Implications and Future Directions

Overall, our results consistently stand in opposition to the Conceptual Leap Hypothesis. In tandem with prior opposing findings (reviewed in the introduction), our work lends strength to alternative theories of inspiration by theorists like Perkins (1983), who argues that conceptual distance does not matter, and Weisberg (2009, 2011), who argues that within-domain expertise is a primary driver of creative cognition. We should be clear that our findings do not imply that *no* creative ideas come from far sources (indeed, in our data, some creative ideas did come from far sources); rather, our data suggest that the most creative design ideas are more likely to come from relying on a preponderance of nearer rather than farther sources. However, our data do suggest that highly creative ideas can often come from relying almost not at all on far sources (as evidenced by the analyses with maximum distance of sources). These good ideas may arise from iterative, deep search, a mechanism for creative breakthroughs that may be often overlooked but potentially at least as important as singular creative leaps (Chan and Schunn 2014; Dow et al. 2009; Mecca and Mumford 2013; Rietzschel et al. 2007; Sawyer 2012; Weisberg 2011). In light of this and our findings, it may be fruitful to deemphasize the privileged role of far sources and mental leaps in theories of design inspiration and creative cognition.

How might this proposed theoretical revision be reconciled with the relatively robust finding that problem-solvers from outside the problem domain can often produce the most creative ideas (Hargadon and Sutton 1997; Franke et al. 2013; Jeppesen 2010)? Returning to our reflections on the potential costs of processing far sources, one way to reconcile the two sets of findings might be to hypothesize that expertise in the distant source domain enables the impact of distant ideas by bypassing the cognitive costs of deeply understanding the far domain, and filters out shallow inferences that are not likely to lead to deep insights. Hargadon and Sutton (1997) findings from their in-depth ethnographic study of the consistently innovative IDEO design firm are consistent with an expertise-mediation claim: the firm's cross-domain-inspired innovations appeared to flow at the day-to-day process level mainly from deep immersion of its designers in multiple disciplines, and "division of expertise" within the firm, with brainstorms acting as crucial catalysts for involving experts from different domains on projects. However, studies directly testing expertise-mediation are scarce or nonexistent.

Further, the weight of the present data, combined with prior studies showing no advantage of far sources, suggests that considering alternative mechanisms of outside-domain advantage may be more theoretically fruitful: for instance, perhaps the advantage of outside-domain problem-solvers arises from the different perspectives they bring to the problem—allowing for more flexible and alternative problem representations, which may lead to breakthrough insights (Knoblich et al. 1999; Kaplan and Simon 1990; Öllinger et al. 2012). Domain outsiders may also have a looser attachment to the *status quo* or prior successful solutions by virtue of being a "newcomer" to the domain (Choi and Levine 2004)—leading to higher readiness to consider good ideas that challenge existing assumptions within the domain—rather than knowledge and transfer of different solutions per se.

Finally, it would be interesting to examine potential moderating influences of source processing *strategies*. In our data, closer sources were more beneficial, but good ideas also did come from far sources; however, as we have argued, it can be more difficult to convert far sources into viable concepts. Are there common strategies for effective conversion of far sources, and are they *different* from strategies for effectively building on near sources? For example, one effective strategy for building on sources while avoiding fixation is to use a schema-based strategy (i.e., extract and transfer abstract functional principles rather than concrete solution features; Ahmed and Christensen 2009; Yu et al. 2014). Are there processing strategies that expert creative designers apply uniquely to far sources (e.g., to deal with potentially un-alignable differences)? Answering this question can shed further light on the variety of ways designers can be inspired by sources to produce creative design ideas.

We close by noting the methodological contribution of this work. While we are not the first to use topic modeling to explore semantic meaning in a large collection of documents, we are the first to our knowledge to validate this method in the context of large-scale study of design ideas. We have shown that the topic model approach adequately captures human intuitions about the semantics of the design space, while providing dramatic savings in cost: indeed, such an approach can make more complex research questions (e.g., exploring pairwise distances between design idea or, tracing conceptual paths/moves in a design ideation session) much more feasible without sacrificing too much quality. We believe this approach can be a potentially valuable way for creativity researchers to study the dynamics of idea generation at scale, while avoiding the (previously inevitable) tradeoff between internal validity (e.g., having adequate statistical power) and external validity (e.g., using real, complex design problems and ideas instead of toy problems).

Appendix 1: Topic Model Technical Details

Document Preprocessing

All documents were first tokenized using the TreeBank Tokenizer from the open-source Natural Language Tool Kit Python library (Bird et al. 2009). To improve the information content of the document text, we removed a standard list of stopwords, i.e., highly frequent words that do not carry semantic meaning on their own (e.g., "the", "this"). We used the open-source MAchine Learning for LanguagE Toolkit's (MALLET; McCallum 2002) stopword list.

Model Parameter Selection

We used MALLET to train our LDA model, with asymmetric priors for the topic-document and topic-word distributions, which allows for some words to be more prominent than others and some topics to be more prominent than others, typically improving model fit and performance (Wallach et al. 2009). Priors were optimized using MALLET's in-package optimization option.

LDA requires that *K* (the number of topics) be prespecified by the modeler. Model fit typically improves with *K*, with diminishing returns past a certain point. Intuitively, higher *K* leads to finer grained topical distinctions, but too high *K* may lead to uninterpretable topics; on the other hand, too low *K* would yield too general topics. Further, traditional methods of optimizing *K* (computing "perplexity", or the likelihood of observing the distribution of words in the corpus given a topic model of the corpus) do not always correlate with human judgments of model quality (e.g., domain expert evaluations of topic quality; Chang et al. 2009).

We explored the following settings of *K*: [12, 25, 50, 100, 200, 300, 400, 500, 600, 700]. Because the optimization algorithm for the prior parameters is nondeterministic, models with identical *K* might produce noticeably different topic model solutions, e.g., if the optimization search space is rugged, the algorithm might get trapped in different local maxima. Therefore, we ran 50 models at each *K*, using identical settings (i.e., 1000 iterations of the Gibbs sampler, internally optimizing parameters for the asymmetric priors). Figure 12.11 shows the mean fit (with both continuous and binary similarity judgments) at each level of *K*.

Model fit is generally fairly high at all levels of *K*, with the continuous judgments tending to increase very slightly with *K*, tapering out past 400. Fit with binary judgments tended to decrease (also very slightly) with *K*, probably reflecting the decreasing utility of increasingly finer grained distinctions for a binary same/different classification. Since we wanted to optimize for fit with human judgments of conceptual distance overall, we selected the level of *K* at which the divergent lines for fit with continuous and binary judgments first begin to cross (i.e., at $K = 400$). Subsequently, we created a combined "fit" measure (sum of the

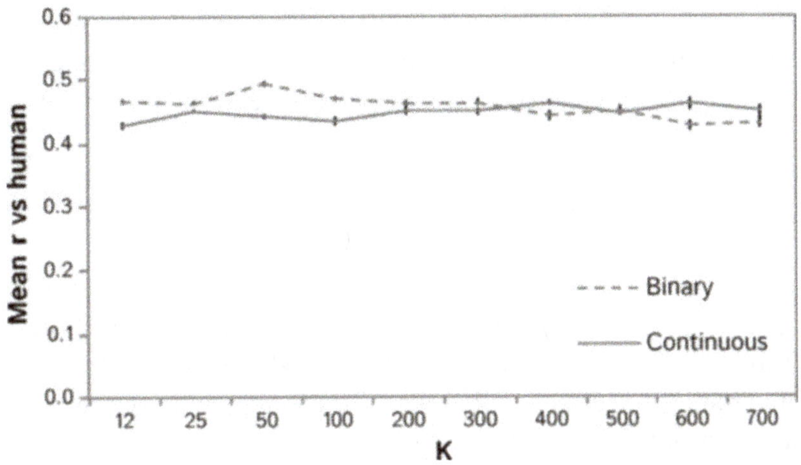

Fig. 12.11 Mean fit (with ± 1 SE) versus human judgments for LDA cosines by level of K

correlation coefficients for fit vs. continuous and binary judgments), and selected the model with $K = 400$ that had the best overall fit measure. However, as we report in the next section, the results of our analyses are robust to different settings of K.

Appendix 2: Statistical Modeling Technical Details

Statistical Modeling Approach

All models were fitted using the lme4 package (Bates et al. 2013) in R (R Core Team 2013), using full maximum likelihood estimation by the Laplace approximation. The following is the general structure of these models (in mixed model notation):

$$\eta_{i(\text{authorj challengek})} = \gamma_{00} + \sum_{q} \gamma_{q0} X_{qi} + u_{0\text{authorj}} + u_{0\text{challengek}}$$

where

- $\eta_{i(\text{author}j\ \text{challenge}k)}$ is the predicted log-odds of being shortlisted for the ith concept posted by the jth author in the kth challenge
- γ_{00} is the grand mean log-odds for all concepts
- γ_{q0} is a vector of q predictors ($q = 0$ for our null model)
- $u_{0\text{authorj}}$ and $u_{0\text{challengek}}$ are the random effects contribution of variation between-authors and between-challenges for mean γ_{00} (i.e., how much a given author or challenge varies from the mean).

A baseline model with only control variables and variance components was first fitted. Then, for the models for both $DIST_{\text{MAX}}$ and $DIST_{\text{MEAN}}$, we first estimated a model with a fixed effect of distance, and then a random effect (to test for problem variation). These random slopes models include the additional parameter $u_{1\text{challenge}k}$ that models the between-challenge variance component for the slope of distance.

Model Selection

Estimates and test statistics for each step in our model-building procedure are shown in Tables 12.3 and 12.4. We first fitted a model predicting Pr (shortlist) with our control variables to serve as a baseline for evaluating the predictive power of our distance measures. The baseline model estimates a strong positive effect of *FEEDBACK*, estimated with high precision: each additional comment added 0.10 [0.07, 0.12] to the log-odds of being shortlisted, $p < .001$. The model also estimated a positive effect of *SHORTSOURCE*, $B = 0.14$ [–0.08, 0.36] but with poor precision, and falling short of conventional statistical significance, $p = 0.21$; nevertheless, we leave it in the model for theoretical reasons. The baseline model is a good fit to the data, reducing deviance from the null model (with no control variables) by a large and statistically significant amount, $\chi^2(1) = 74.35$, $p = .00$.

For the fixed slope model for $DIST_{MAZ}$, adding the coefficient for results in a significant reduction in deviance from the baseline model, $\chi^2(2) = 0.13$, $p = .47$.

Table 12.4 Model estimates and fit statistics for cross-classified multilevel logistic regressions of Pr (shortlist) on $DIST_{MAX}$, with comparison to baseline model (controls only)

	Baseline model (controls only)	$DIST_{MAX}$, fixed slope	$DIST_{MAX}$, random slope
Fixed effects			
γ_{00}, intercept	$-2.66^{[-3.28,\ -2.03]}$	$-2.57^{[-3.29,\ -2.05]}$	$-2.57^{[-3.29,\ -2.05]}$
γ_{10}, *FEEDBACK*	$0.09^{***[0.07,\ 0.12]}$	$0.10^{***[0.07,\ 0.12]}$	$0.10^{***[0.07,\ 0.12]}$
γ_{20}, *SOURCESHORT*	$0.14^{[-0.08,\ 0.36]}$	$0.15^{[-0.07,\ 0.38]}$	$0.15^{[-0.07,\ 0.38]}$
γ_{30}, $DIST_{MAX}$		$-0.33^{*[-0.60,\ -0.06]}$	$-0.32^{*[-0.59,\ -0.06]}$
Random effects			
$u_{0\text{author}j}$ for intercept	0.29	0.31	0.32
$u_{0\text{challenge}k}$ for intercept	0.75	0.76	0.74
$u_{3\text{challenge}k}$ for $DIST_{MAX}$			0.00
Model fit statistics			
Deviance	511.39	506.04	505.99
AIC	521.39	518.04	521.99

$^{m}p < .10$, $^{*}p < .05$, $^{**}p < .01$, $^{***}p < .001$, 95% CI (Wald) = [lower, upper]

The random slope model did not significantly reduce deviance in comparison with the simpler fixed slope model, $\chi^2(2) = 0.05$, $p = .49$ (p-value is halved, heeding common warnings that a likelihood ratio test discriminating two models that differ on only one variance component may be overly conservative, e.g., Pinheiro and Bates 2000). Also, the Akaike Information Criterion (AIC) increases from the fixed to random slope model. Thus, we select the fixed slope model (i.e., no problem variation) as our best estimate of the effects of $DIST_{MAX}$. This final model has an overall deviance reduction versus null at $\chi^2(3) = 79.71$, $p = .00$.

We used the same procedure for model selection for the $DIST_{MEAN}$ models. The fixed slope model results in a small but significant reduction in deviance from the baseline model, $\chi^2(1) = 6.27, p = .01$. Adding the variance component for the slope of $DIST_{MEAN}$ increases the AIC, and does not significantly reduce deviance, $\chi^2(2) = 0.07$, $p = .48$ (again, p-value here is halved to correct for overconservativeness). Thus, again we select the fixed slope model as our final model for the effects of $DIST_{MEAN}$. This final model has an overall reduction in deviance from the null model of about $\chi^2(3) = 80.61$, $p = .00$ (Table 12.5).

Robustness and Sensitivity

We tested the robustness of our coefficient estimates by calculating outlier influence statistics using the influence.measures method in the stats package in R, applied to

Table 12.5 Model estimates and fit statistics for cross-classified multilevel logistic regressions of Pr (shortlist) on $DIST_{MEAN}$, with comparison to baseline model (controls only)

	Baseline model (controls only)	$DIST_{MEAN}$, fixed slope	$DIST_{MEAN}$, random slope
Fixed effects			
γ_{00}, intercept	$-2.66^{[-3.28,\ -2.03]}$	$-2.74^{[-3.36,\ -2.11]}$	$-2.74^{[-3.36,\ -2.11]}$
γ_{10}, *FEEDBACK*	$0.09^{***[0.07,\ 0.12]}$	$0.10^{***[0.07,\ 0.12]}$	$0.10^{***[0.07,\ 0.12]}$
γ_{20}, *SOURCESHORT*	$0.14^{[-0.08,\ 0.36]}$	$0.13^{[-0.09,\ 0.35]}$	$0.13^{[-0.09,\ 0.35]}$
γ_{30}, $DIST_{MEAN}$		$-0.40^{*[-0.71,\ -0.09]}$	$-0.40^{*[-0.73,\ -0.07]}$
Random effects			
$u_{0\text{author}j}$ for intercept	0.29	0.31	0.30
$u_{0\text{challenge}k}$ for intercept	0.75	0.73	0.73
$u_{3\text{challenge}k}$ for $DIST_{MEAN}$			0.03
Model fit statistics			
Deviance	511.39	505.13	505.06
AIC	521.39	517.13	521.06

$^m p < .10$, $^*p < .05$, $^{**}p < .01$, $^{***}p < .001$, 95% CI (Wald) = [lower, upper]

logistic regression model variants of both the $DIST_{MEAN}$ and $DIST_{MAX}$ models (i.e., without author- and challenge-level variance components; coefficient estimates are almost identical to the fixed slope multilevel models): DFBETAS and Cook's Distance measures were below recommended thresholds for all data points (Fox 2002).

Addressing potential concerns about sensitivity to topic model parameter settings, we also fitted the same fixed slope multilevel models using recomputed conceptual distance measures for the top 20 (best-fitting) topic models at $K = 200$, 300, 400, 500, and 600 (total of 100 models). All models produced negative estimates for the effect of both $DIST_{MEAN}$ and $DIST_{MAX}$, with poorer precision for lower K. Thus, our results are robust to different settings of K for the topic models.

We also address potential concerns about interactions with expertise by fitting a model that allowed the slope of distance to vary by authors. In this model, the overall mean effect of distance remained almost identical ($B = -0.46$), and the model's fit was not significantly better than the fixed slope model, $\chi^2(3) = 3.44$, $p = .16$, indicating a lack of statistically significant between-author variability for the slope of distance.

Finally, we also fitted models that considered not just immediately cited inspirations, but also indirectly cited inspirations (i.e., inspirations cited by cited inspirations), and they too yielded almost identical coefficient estimates and confidence intervals.

References

Amabile, T. M. (1982). Social psychology of creativity: A consensual assessment technique. *Journal of Personality and Social Psychology, 43*(5), 997–1013.

Ahmed, S., & Christensen, B. T. (2009). An in situ study of analogical reasoning in novice and experienced designer engineers. *Journal of Mechanical Design, 131*(11), 111004.

Baer, J., & McKool, S. S. (2009). Assessing creativity using the consensual assessment technique. In C. S. Schreiner (Ed.), *Handbook of research on assessment technologies, methods, and applications in higher education* (pp. 65–77). PA: Hershey.

Bates, D., Maechler, M., Bolker, B., & Walker, S. (2013). *Lme4: Linear mixed-effects models using Eigen and S4. R package version 1.0-5* [Computer Software]. Retrieved from http://CRAN.R-project.org/package=lme4.

Bird, S., Klein, E., & Loper, E. (2009). *Natural language processing with python*. Sebastopol, CA: O'Reilly Media Inc.

Blei, D. M. (2012). Probabilistic topic models. *Communications of the ACM, 55*(4), 77–84.

Blei, D. M., & Lafferty, J. D. (2006). Dynamic topic models. In *Proceedings of the 23rd international conference on machine learning* (pp. 113–120).

Blei, D. M., & Lafferty, J. D. (2007). A correlated topic model of science. *The Annals of Applied Statistics, 1,* 17–35.

Blei, D. M., Ng, A. Y., Jordan, M. I., & Lafferty, J. (2003). Latent dirichlet allocation. *Journal of Machine Learning Research, 3,* 993–1022.

Brown, A. S., & Murphy, D. R. (1989). Cryptomnesia: Delineating inadvertent plagiarism. *Journal of Experimental Psychology: Learning, Memory, and Cognition, 15*(3), 432–442.

Brown, R. (1989). Creativity: What are we to measure? In J. A. Glover, R. R. Ronning, & C. R. Reynolds (Eds.), *Handbook of creativity* (pp. 3–32). New York, NY: Plenum Press.

Button, K. S., Ioannidis, J. P. A., Mokrysz, C., Nosek, B. A., Flint, J., Robinson, E. S. J., et al. (2013). Power failure: Why small sample size undermines the reliability of neuroscience. *Nature Reviews Neuroscience, 14*(5), 365–376. https://doi.org/10.1038/nrn3475.

Chakrabarti, A. (2006). Defining and supporting design creativity. In *Proceedings of the 9th international design conference DESIGN 2006* (pp. 479–486).

Chan, J., & Schunn, C. (2014). The impact of analogies on creative concept generation: Lessons from an in vivo study in engineering design. *Cognitive Science, 39*(1), 126–155.

Chan, J., Fu, K., Schunn, C. D., Cagan, J., Wood, K. L., & Kotovsky, K. (2011). On the benefits and pitfalls of analogies for innovative design: Ideation performance based on analogical distance, commonness, and modality of examples. *Journal of Mechanical Design, 133,* 081004.

Chang, J., Gerrish, S., Wang, C., Boyd-graber, J. L., & Blei, D. M. (2009). Reading tea leaves: How humans interpret topic models. In *Advances in neural information processing systems* (pp. 288–296).

Choi, H. S., & Levine, J. M. (2004). Minority influence in work teams: The impact of newcomers. *Journal of Experimental Social Psychology, 40*(2), 273–280.

Chiu, I., & Shu, H. (2012). Investigating effects of oppositely related semantic stimuli on design concept creativity. *Journal of Engineering Design, 23*(4), 271–296. https://doi.org/10.1080/09544828.2011.603298.

Dahl, D. W., & Moreau, P. (2002). The influence and value of analogical thinking during new product ideation. *Journal of Marketing Research, 39*(1), 47–60.

Deerwester, S., Dumais, S. T., Furnas, G. W., & Landauer, T. K. (1990). Indexing by latent semantic analysis. *Journal of the American Society for Information Science, 41*(6), 1990.

Dow, S. P., Heddleston, K., & Klemmer, S. R. (2009). The efficacy of prototyping under time constraints. In *Proceedings of the 7th ACM conference on creativity and cognition.*

Dunbar, K. N. (1997). How scientists think: On-line creativity and conceptual change in science. In T. B. Ward, S. M. Smith, & J. Vaid (Eds.), *Creative thought: An investigation of conceptual structures and processes* (pp. 461–493). Washington, DC: American Psychological Association Press.

Dym, C. L. (1994). *Engineering design: A synthesis of views.* New York, NY: Cambridge University Press.

Eckert, C., & Stacey, M. (1998). Fortune favours only the prepared mind: Why sources of inspiration are essential for continuing creativity. *Creativity and Innovation Management, 7*(1), 1–12.

Enkel, E., & Gassmann, O. (2010). Creative imitation: Exploring the case of cross-industry innovation. *R & D Management, 40*(3), 256–270.

Findlay, A. (1965). *A hundred years of chemistry* (3rd ed.). London: Duckworth.

Fox, J. (2002). *An R and s-plus companion to applied regression.* Thousand Oaks, CA: Sage.

Franke, N., Poetz, M. K., & Schreier, M. (2013). Integrating problem solvers from analogous markets in new product ideation. *Management Science, 60*(4), 1063–1081.

Freeman, A., & Golden, B. (1997). *Why didn't I think of that? Bizarre origins of ingenious inventions we couldn't live without.* New York: Wiley.

Fu, K., Chan, J., Cagan, J., Kotovsky, K., Schunn, C., & Wood, K. (2013). The meaning of "near" and "far": The impact of structuring design databases and the effect of distance of analogy on design output. *Journal of Mechanical Design, 135*(2), 021007. https://doi.org/10.1115/1.4023158.

Gentner, D., & Markman, A. B. (1997). Structure mapping in analogy and similarity. *American Psychologist, 52*(1), 45–56.

German, T. P., & Barrett, H. C. (2005). Functional fixedness in a technologically sparse culture. *Psychological Science, 16*(1), 1–5.

Gero, J. S. (2000). Computational models of innovative and creative design processes. *Technological Forecasting and Social Change, 64*(2), 183–196.

Goldschmidt, G., & Smolkov, M. (2006). Variances in the impact of visual stimuli on design problem solving performance. *Design Studies, 27*(5), 549–569.

Gonçalves, M., Cardoso, C., & Badke-Schaub, P. (2013). Inspiration peak: Exploring the semantic distance between design problem and textual inspirational stimuli. *International Journal of Design Creativity and Innovation, 1*(ahead-of-print), 1–18.

Griffiths, T. L., & Steyvers, M. (2004). Finding scientific topics. *Proceedings of the National academy of Sciences of the United States of America, 101*(Suppl 1), 5228–5235. https://doi.org/10.1073/pnas.0307752101.

Hargadon, A., & Sutton, R. I. (1997). Technology brokering and innovation in a product development firm. *Administrative Science Quarterly, 42*(4), 716. https://doi.org/10.2307/2393655.

Helms, M., Vattam, S. S., & Goel, A. K. (2009). Biologically inspired design: Process and products. *Design Studies, 30*(5), 606–622.

Hender, J. M., Dean, D. L., Rodgers, T. L., & Jay, F. F. (2002). An examination of the impact of stimuli type and GSS structure on creativity: Brainstorming versus non-brainstorming techniques in a GSS environment. *Journal of Management Information Systems, 18*(4), 59–85.

Holyoak, K. J., & Thagard, P. (1996). *Mental leaps: Analogy in creative thought*. Cambridge, MA: MIT press.

Impact Stories. (n.d.). *Impact stories*. [Web page] Retrieved from http://www.openideo.com/content/impact-stories.

Jansson, D. G., & Smith, S. M. (1991). Design fixation. *Design Studies, 12*(1), 3–11.

Jeppesen, L. B., & Lakhani, K. R. (2010). Marginality and problem-solving effectiveness in broadcast search. Organization Science, *21*(5), 1016e1033.

Jessup, E. R., & Martin, J. H. (2001). Taking a new look at the latent semantic analysis approach to information retrieval. In *Computational information retrieval* (pp. 121–144). Philadelphia: SIAM.

Kaplan, C., & Simon, H. A. (1990). In search of insight. *Cognitive Psychology, 22*(3), 374–419.

Kalogerakis, K., Lu, C., & Herstatt, C. (2010). Developing innovations based on analogies: Experience from design and engineering consultants. *Journal of Product Innovation Management, 27,* 418–436.

Knoblich, G., Ohlsson, S., Haider, H., & Rhenius, D. (1999). Constraint relaxation and chunk decomposition in insight problem solving. *Journal of Experimental Psychology. Learning, Memory, and Cognition, 25*(6), 1534–1555.

Liikkanen, L. A., & Perttula, M. (2008). Inspiring design idea generation: Insights from a memory-search perspective. *Journal of Engineering Design, 21*(5), 545–560.

Linsey, J., Tseng, I., Fu, K., Cagan, J., Wood, K., & Schunn, C. (2010). A study of design fixation, its mitigation and perception in engineering design faculty. *Journal of Mechanical Design, 132* (4), 0410031–04100312.

Malaga, R. A. (2000). The effect of stimulus modes and associative distance in individual creativity support systems. *Decision Support Systems, 29*(2), 125–141.

Marsh, R. L., Landau, J. D., & Hicks, J. L. (1997). Contributions of inadequate source monitoring to unconscious plagiarism during idea generation. *Journal of Experimental Psychology: Learning, Memory, and Cognition, 23*(4), 886–897.

Marsh, R. L., Ward, T. B., & Landau, J. D. (1999). The inadvertent use of prior knowledge in a generative cognitive task. *Memory and Cognition, 27*(1), 94–105.

McCallum, A. K. (2002). *MALLET: A machine learning for language toolkit.* [Computer Software] Retrieved from http://mallet.cs.umass.edu.

Mecca, J. T., & Mumford, M. D. (2013). Imitation and creativity: Beneficial effects of propulsion strategies and specificity. *The Journal of Creative Behavior, 48*(3), 209–236. https://doi.org/10.1002/jocb.49.

Öllinger, M., Jones, G., Faber, A. H., & Knoblich, G. (2012). Cognitive mechanisms of insight: The role of heuristics and representational change in solving the eight-coin problem. *Journal of Experimental Psychology: Learning, Memory, and Cognition, 39*(3), 931. https://doi.org/10.1037/a0029194.

Perkins, D. N. (1983). Novel remote analogies seldom contribute to discovery. *The Journal of Creative Behavior, 17*(4), 223–239.

Perkins, D. N. (1997). Creativity's camel: The role of analogy in invention. In T. B. Ward, S. M. Smith, & J. Vaid (Eds.), *Creative thought: An investigation of conceptual structures and processes* (pp. 523–538). Washington D.C.: American Psychological Association.

Pinheiro, J. C., & Bates, D. M. (2000). *Linear mixed-effects models: Basic concepts and examples*. Berlin: Springer.

Poze, T. (1983). Analogical connections: The essence of creativity. *The Journal of Creative Behavior, 17*(4), 240–258.

Raudenbush, S. W., & Bryk, A. S. (2002). *Hierarchical linear models: Applications and data analysis methods* (2nd ed.). CA: Thousand Oaks.

R Core Team. (2013). *R: A language and environment for statistical computing* [Computer Software]. Vienna, Austria: R Foundation for Statistical Computing. Retrieved from http://www.R-project.org/.

Rietzschel, E. F., Nijstad, B. A., & Stroebe, W. (2007). Relative accessibility of domain knowledge and creativity: The effects of knowledge activation on the quantity and originality of generated ideas. *Journal of Experimental Social Psychology, 43*(6), 933–946.

Sawyer, R. K. (2012). *Explaining creativity: The science of human innovation* (2nd ed.). New York: Oxford University Press.

Schwartz, H. A., Eichstaedt, J. C., Kern, J. K., Dziurzynski, J. K., Ramones, M. L. D., Agrawal, L. R., et al. (2013). Personality, gender, and age in the language of social media: The open-vocabulary approach. *PLoS ONE, 8*(9), e73791.

Schunn, C. D., & Dunbar, K. N. (1996). Priming, analogy, and awareness in complex reasoning. *Memory and Cognition, 24*(3), 271–284.

Steyvers, M., & Griffiths, T. (2007). Probabilistic topic models. In T. Landauer, D. McNamara, S. Dennis, & W. Kintsch (Eds.), *Handbook of latent semantic analysis* (pp. 424–440). New York, NY: Lawrence Erlbaum.

Tseng, I., Moss, J., Cagan, J., & Kotovsky, K. (2008). The role of timing and analogical similarity in the stimulation of idea generation in design. *Design Studies, 29*(3), 203–221.

Ullman, D. (2002). *The mechanical design process* (3rd ed.). New York: McGraw Hill.

Wallach, H. M., Mimno, D. M., & McCallum, A. (2009). Rethinking LDA: Why priors matter. *Neural Information Processing Systems, 22,* 1973–1981.

Ward, T. B. (1994). Structured imagination: The role of category structure in exemplar generation. *Cognitive Psychology, 27*(1), 1–40.

Ward, T. B. (1998). Analogical distance and purpose in creative thought: Mental leaps versus mental hops. In K. J. Holyoak, D. Gentner, & B. Kokinov (Eds.), *Advances in analogy research: Integration of theory and data from the cognitive, computational, and neural sciences* (pp. 221–230). Bulgaria: Sofia.

Weisberg, R. W. (2009). On "out-of-the-box" thinking in creativity. In A. B. Markman & K. L. Wood (Eds.), *Tools for innovation* (pp. 23–47). NY: New York.

Weisberg, R. W. (2011). Frank lloyd wright's fallingwater: A case study in inside-the-box creativity. *Creativity Research Journal, 23*(4), 296–312. https://doi.org/10.1080/10400419.2011.621814.

Wiley, J. (1998). Expertise as mental set: The effects of domain knowledge in creative problem solving. *Memory and Cognition, 26*(4), 716–730.

Wilson, J. O., Rosen, D., Nelson, B. A., & Yen, J. (2010). The effects of biological examples in idea generation. *Design Studies, 31*(2), 169–186.

Yu, L., Kraut, B., and Kittur, A. (2014). Distributed analogical idea generation: innovating with crowds. In *Proceedings of the ACM Conference on Human Factors in Computing Systems (CHI'14)*.

Zeger, S. L., Liang, K.-Y., & Albert, P. S. (1988). Models for longitudinal data: A generalized estimating equation approach. *Biometrics, 44,* 1049–1060.

Chapter 13
Integrating is Caring? Or, Caring for Nanotechnology? Being an Integrated Social Scientist

Ana Viseu

13.1 Introduction[1]

In the Western world, governments and funding bodies increasingly emphasize the need for "responsible research and innovation" practices the integration of the social sciences and humanities within publicly funded research and development (R&D) initiatives as a means to foster "responsible innovation" (Public Law 108-153; European Commission 2004, 2014). Although the concept "responsible innovation" remains vague, it tends to assume the integration of the social sciences within research and development (R&D) initiatives, with the goal of maximizing societal benefits while reducing the possibility of negative impacts and public controversy. In the landscape of integration, nanotechnology[2] has emerged as a primary field site, and the institutional hiring social scientists is one of its instruments. Integrated social scientists are asked to *care* for nanotechnology research and development by learning how to observe but not disturb. They are assumed to

[1]This chapter is a re-reprint of an article published in 2015 (Viseu, 2015a). In preparation for this new publication, I made some minor changes and updated the literature and bibliographic sources. At the end of the chapter, I also include a new section that reflects on the three years that have gone by since the printing of the original article.

[2]In many ways, the term 'nanotechnology' is equally vague and different actors enact it differently. The National Nanotechnology Initiative (NNI) describes it as "science, engineering, and technology conducted at the nanoscale, which is about 1–100 nm" (NNI, n.d). In other words, nanotechnology is encompassing referring to the study, experimentation, tools, and knowledge of matter at the nanoscale.

A. Viseu (✉)
Universidade Europeia, Lisbon, Portugal
e-mail: ana@anaviseu.org

A. Viseu
Centro Interuniversitário de História das Ciências e Tecnologia,
Universidade de Lisboa, Lisbon, Portugal

E. Subrahmanian et al. (eds.), *Engineering a Better Future*,
https://doi.org/10.1007/978-3-319-91134-2_13

broaden the field, while simultaneously guarding its boundaries, shielding it from that which (and those who) stand outside "proper" science. Care emerges as a governance strategy, with social scientists being cast as the main caregivers.

At the core of this chapter is an attempt to engage empirically, conceptually, and affectively with the cares of integration in nanotechnology from the standpoint of the *integrated* social scientist. To follow integration's journey from policy to embodiment, I draw upon my three-year experience as the in-house social scientist at the Cornell NanoScale Facility (CNF) and the United States' National Nanotechnology Infrastructure Network (NNIN). I argue that, despite integration's potential for creating new modes of collaboration between the social and natural sciences, it is based on traditional and prescriptive arrangements, where disciplinary boundaries, funding provisions and power asymmetries are reified not challenged. Given this socio-material arrangement, integration's potential for opening up science to distinct modes of being, doing and is muted, leading me to conclude that, from the standpoint of those working within Science and Technology Studies (STS), the institutionalized integration of social scientists within R&D initiatives is designed to fail insofar as social scientists are being called upon to care for nanotechnology by keeping it undisturbed. Moreover, when failure is by design a personal failure—entangled with one's identity and career trajectory—the affective cost of caring for nanotechnology is exceptionally high.

Written in the first person, this article is not meant as a 'confessional tale' (Van Maanen, 1988: 74) that expiates my data generation or fieldwork, nor is it an exercise in reflexivity that conjures me as the holder of a privileged 'epistemic, moral or political virtue' (Lynch, 2000) that my colleagues and subjects did not possess. My goal is to use my three-year experience enacting this policy to write up and analyze the socio-material culture of integration. It is a personal story that puts me at the center of the narrative, and in so doing it is an uncomfortable story, one that I have previously resisted putting into print, even while sharing it at conferences and in lectures. In the course of these events, I heard back from many others (mainly women) who relayed accounts of similar situations, and to their silent voices I add my own. Describing my personal experience is then a means to counter the strategy of individualizing problems in order to turn a blind eye to systemic and collective failures. This is reason enough to write this chapter, but there are others. Examining the current enactments of integration policy is important for Science and Technology Studies (STS) - as well as scholars interested in science and science policy - because integration is going mainstream, slowly becoming a preferred policy tool for the study and management of science and its relations with society; because it effects changes to STS' methods, identity, sustainability, and research sites—helping shape what constitutes a research subject/object, the researcher's position within it, and the kinds of knowledge that can be generated; because it is being done in STS' name[3]; and, because, in its current enactment, integration is of

[3]It is hard to know exactly how much money is being allocated to 'societal dimensions' research. For instance, National Nanotechnology Initiative (NNI) claims to have spent US$182.2 million of

dubious value and is gaining traction without adequate examination. Finally, this analysis is important because conducting it will help to move us towards an analytically skeptical (but not wholly dismissive) perspective on integration, which is key to the development of more suitable collaboration and governance models.

13.2 Why Care?

The demand to integrate social sciences issues and concerns and scientific practice and agendas is neither new nor exclusive to nanotechnology. The emergence of demand for integration can be traced back to the 1980s, when a number of trends combined to challenge the established way of publicly funding science. Political demands for greater economic "return-on-investment", public questioning of the direction of technoscientific development, as well as a waning of confidence in the self-regulation of science, all led to a renegotiation of relations between science, the state and citizens (Gibbons 1999; Gibbons et al. 1994; Nowotny et al. 2001; Jasanoff 2005, 2011). Created in 1988, the Human Genome Project's (HGP) "Ethical, Legal, and Societal Implications" (ELSI) program emerged as the first materialization of a model of scientific governance that formally sought to enroll the social sciences and humanities within its midst. It was against the backdrop of the ELSI program, and critiques of its lack of independence, power, and mandate (Wolfe 2000; Fisher 2005; McCain 2002), that nanotechnology policy was brought to life.

The U.S. National Nanotechnology Initiative (NNI) was created in the year 2000 by the Clinton administration to "expedite the discovery, development and deployment of nanoscale science and technology to serve the public good" (NNI, n. d). A total of three years later, the U.S. Congress passed legislation to regulate the NNI, stipulating that all NNI funded research should "insofar as possible, *integrate* research [and activities] on societal, ethical, and environmental concerns with nanotechnology research and development" (Public Law 108-153, 2003, my emphasis). Albeit pointing to integration as the path forward, the law remains mute on the practical and important issue of how such research will be identified, defined, managed, sponsored, or implemented. Roco and Bainbridge (2001)—two of the architects behind the creation of the NNI–described their rationale for integrating the social sciences and humanities in nano:

> The inclusion of social scientists and humanistic scholars, such as philosophers of ethics, in the social process of setting visions for nanotechnology is an important step for the NNI. As scientists or dedicated scholars in their own right, they can respect the professional integrity of nanoscientists and nanotechnologists, while contributing a fresh perspective. Given

its budget on 'education and social dimensions' from 2006 to 2010. While this is a lot of money, Guston (2010) clarifies that it amounts to 2.3% of the total budget, which is less than the reputed 3–5% dedicated to Ethical, Legal and Societal Implications (ELSI) during the Human Genome Program. Moreover, Guston argues that most of this funding is for educational endeavors and only between 0.5 and 1% is dedicated to funding 'societal dimensions' activities.

> appropriate support, they could inform themselves deeply enough about a particular nanotechnology to have a well-grounded evaluation. At the same time, they are professionally trained representatives of the public interest and capable of functioning as communicators between nanotechnologists and the public or government officials. Their input may help maximize the societal benefits of the technology while reducing the possibility of debilitating public controversies. (Roco and Bainbridge 2001: 12)

In other words, social scientists and humanities scholars (narrowly characterized as those studying ethics) are called upon because, being scientists themselves, they are more likely (than the public, one assumes) to understand and be respectful of professional and disciplinary boundaries. To ensure that these boundaries are not crossed, social scientists are summoned to participate in the "social process of setting visions" for nanotechnology, with a few—those who prove themselves willing to being educated in the correct scientific facts—being allowed to go on to provide "well-grounded evaluations" (i.e., based on the aforementioned proper education) of specific nanotechnologies. Finally, social scientists are depicted as communicators and as specialists in the (homogeneous) public, able to contribute to the smooth development and progress of nanotechnology. In sum, integration is imagined less as a means to blur boundaries and extend or open the field of nanotechnology, than as a means to maintain the status quo. Integration reifies a prevailing "turn to ethics" (Jasanoff 2011) that splits the study of technoscience into a domain of values (deemed subjective, extrascientific, and falling within the scope of the social sciences), and a domain of facts (seen as objective, universal, and for experts only). In the integration framework depicted here, social scientists and humanities scholars are entrusted with the job of *caring for* nanotechnology by learning how to observe, protect, and communicate, without disturbing technical or professional boundaries.

A number of scholars, most prominently feminist science scholars, have long argued that making care visible—both as a practice and an ethos—is key to the study of science and technology (Fox 1980; Fox Keller 1987; Armstrong and Armstrong 2002). Caring, these authors remind us, is not only hard work, it is also work that is often seen as feminine and affect-oriented, work that is devalued and made invisible in the worlds of science and technology. Care is devalued because it is feminized and vice versa. As of late, care has again become front and center in STS studies, namely with Puig de la Bellacasa's (2011) call for science studies scholars to attend to "matters of care" in their engagements with the world-making practices of technoscience (see also, Mol et al. 2010). Matters of care, Puig de la Bellacasa argues, speak of knowledge as doing and intervening, that is, of the ethico-political dimensions of knowledge. They do not replace Latour's (2004) "matters of concern" but rather direct our attention to the affective dimension of our concerns. They do so by evidencing the worries, cares, and responsibilities that are constitutive not only of our engagements, but also of the very things and people we study, and worlds we co-create.

Yet, in the context of integration, care follows a traditional, prescriptive approach emerging as a reductive concept and governance strategy. This richness of contrast, and not the fact that policymakers or scientists use the term, is one reason why "care" is an important heuristic for the study of integration. Examining integration through the lens of care illustrates the multiple definitions, dimensions, prescriptions, and affordances of integration and of being an integrated social scientist. Moreover, care, in the words of Haraway (2010), allows me to "stay with the trouble" (see also, Murphy 2015) of integration in nanotechnology, both by troubling and being troubled by it. In sum, examining integration as a kind of care practice *and* as a matter of care allows me to highlight its often invisible, existential, epistemic, and affective costs, as well as its practices and enactments.

Elsewhere, I have examined the workings of the integration model from the vantage point of nanotechnology practitioners (Viseu and Maguire 2012). Focusing on how practitioners integrate the "social and ethical dimensions" of nanotechnology into their practices, we conclude that current policy has succeeded at normalizing a kind of inclusion of social and ethical issues in the landscape of nanotechnology. This inclusion confers obligations to be fulfilled mostly through outreach and education, as well as personal/moral choices around what (not) to study. At the same time, as some authors have argued, integration has been less than successful at fostering modes of thought, research ethos, or joint relationalities that are inclusive of commitments to social and ethical issues. This chapter extends such arguments by providing evidence of how narrow conceptualizations of the social and the ethical then prompt particular expectations of what an in-house social scientist can, and should, do.

The social sciences have not been passively cast as caretakers in the nanotechnology effort. Instead, they have had a central role in making nanotechnology an issue worthy of care.[4] Making this point, Nordmann (2007) asserts that the work of maintaining nanotechnology's reality is conducted not only "by advocates and activists, visionary policy makers, [and] scientists when they speak to the public or argue for future funding" but also by "philosophers, ethicists, and social scientists" (Nordmann 2007: 223). It is the latter, he argues, who "have… been recruited to do some of the work that is required to convince a larger audience that 'nanotechnology'… [is a] meaningful concept" (Nordmann 2007: 223). The two centers for "nanotechnology in society" created in 2005 attest to this, as do the 308 nanotechnology-related social science articles published between 1998 and 2007 (Shapira et al. 2010), and the "Society for the Study of Nanoscience and Emerging Technologies" created in 2007, to name a few. All these engagements are both products and agents in making nanotechnology into a topic that deserves to be cared for (see also, Karinen and Guston 2010).

STS both mirrors and drives this shift with its growing interest in "engaged" or "interventionist" research. Recall, for instance, the widely cited discussion in the pages of *Science, Technology & Human Values* between Webster (2007) and Wynne (2007) on the relationship of STS to science policy, and its possible future as a "serviceable" discipline. STS, one can argue, is often engaged with the worlds it studies (Sismondo 2008); but here I am discussing a particular kind of

[4]A colleague who participated in Europe's Deepening Ethical Engagement and Participation with Emerging Nanotechnologies (DEEPEN) told me that one of the project's main conclusions was that it had created a public for nano (João Arriscado Nunes 2011, personal communication; see also, Laurent 2017).

engagement: that which calls for the institutionalized and formal hiring of social scientists in technical facilities.[5] Described in the literature as "embedded humanists" (Schuubiers and Fisher 2009), "convergence workers" (Gannon 2009), or "engagement agents" (Te Kulve and Rip 2011), integrated scholars are assumed to move closer to the decision-making locus, no longer critical observers but more influential insider agents and/or policy advisors (Webster 2007; Rip 2006, 2009; Stegmaier 2009; Schuurbiers 2011). Yet, reports by in-house social scientists remain mostly ambivalent (Calvert and Martin 2009; Gorman 2011; Hackett and Rhoten 2011; Schuurbiers 2011; Balmer et al. 2012; Doubleday and Viseu 2010; Suchman 2013; Klenk and Meehan 2015; Fitzgerald et al. 2014). And these are the success stories. Experiences of failure, that is, experiences deemed by the embedded social scientist to have failed to create new forms of collaborative scientific practice, are mostly not published, being instead reported at conferences where they are voiced with some shame and received with enthusiasm (Viseu 2012; Aguiton 2012; Thoreau 2012).[6] The silence surrounding integration attests to its affective cost, one that is compounded by the integrated researcher's position as an insider (embraced as friendly "caretaker"), and an outsider (feared as a "critic"). On this matter, Hackett and Rhoten (2011) conclude that no matter how committed to advancing knowledge an integrated researcher is, after spending a number of years working within an organization, she will attempt to avoid "any possibility" of causing harm. It is not a question, they say, of "going native"; instead it is related to the "liminal status" occupied by these researchers and the conflicts of values inherent to their precarious position as insider/outsider (Hackett and Rhoten 2011: 835; see also, Balmer et al. 2012; Stegmaier 2009).

A few voices have started to speak to the paradoxes of integration. Rabinow and Bennett (2012) examine their integration at the National Science Foundation (NSF) funded "Synthetic Biology Engineering Research Center" (SynBERC), and their failure to "invent a new form of ethical practice" (p. 5) where natural and social scientists would collaborate as equals. The authors depict an environment of deep-seated and untouchable power asymmetries, characterized by resistance both to change and to learning new modes of (collaborative) doing. They portray a politics of funding that made them the only actors exclusively dependent on money from SynBERC, and indicate a confluence between the natural scientists' and the funders' traditional understanding of ELSI as ethics, which functioned to the detriment of the authors' vision (see also, Balmer et al. 2012; Doubleday and Viseu 2010; Viseu and Maguire 2012). Their integration experiment concluded with Rabinow's dismissal, and Rabinow and Bennett's (2012) subsequent realization

[5]Two other strategies—the creation of centers dedicated to the study of nanotechnology's societal, ethical and legal implications, and the sponsoring of outreach and educational activities within and by NNI funded technical centers—remain beyond the scope of this chapter.

[6]Anthropologist Christopher Kelty University of California, Los Angeles, (UCLA)) has received so many emails from (mainly female) social scientists concerned about their embeddedness that he has started a database with their names (Christopher Kelty, January 13, 2012, personal communication).

that they "should have known better", that theirs was a Faustian bargain where they "underestimated the existential price to be paid" (Rabinow and Bennett 2012: 173).

Despite its ambivalent track record, integration is increasingly used as a preferred policy tool and as a model for many of STS engagements with technoscience, making it all the more important to examine whether and how it is working. Using care as a heuristic enables me to follow the above cited "existential price" of being integrated, as well as the instructions and articulations that make it pricey. In the next section, I draw upon my experience as the in-house social scientist at a nanofabrication network and facility to address the following questions: What are the parameters of integration? What modes of work and caring are included and excluded? How are the roles of social scientists as care workers defined? What kinds of knowledge are afforded and prevented within this arrangement? What are the costs of being integrated? What constitutes success, for whom and at what cost? My account revolves around the following main issues: the (unacknowledged) different goals of the actors involved; deep asymmetries in power, funding, and personnel numbers; the undervalued status of the social scientist's identity, skills, and knowledge within the initiative; lack of independence; and, finally, affective and epistemic costs of integration. As stated earlier, the details of my experience are intrinsically personal, but they are supported by accounts shared with me by other researchers in similar positions.

13.3 Caring for Nanotechnology

Despite integration's novelty as a policy framework, it often comes into being through a top-down request from a funding body, no different from requiring that grant applications include a section on "broader impacts" or "societal value" (see Rabinow and Bennett 2012; Karinen and Guston 2010). Take, for example the case of the 2003 NSF's solicitation for the creation of a network of user facilities, a "National Nanotechnology Infrastructure Network" (NNIN), to support the United States' future infrastructure needs for research and education at the nanoscale. The 16-page solicitation described in loose terms the need to consider the "social and ethical implications" of nanotechnology as "additional review criteria" (NSF 2003:12), requesting that applications "list the issues that will be core concerns… [and] describe plans to facilitate … cooperation and interchange between scientists and engineers in nanoscience-related fields and social scientists and ethicists studying nanotechnology" (NSF 2003: 9).

One of the networks applying for the NNIN grant was led by the Cornell Nanoscale Facility (CNF), and included among its team Professor Bruce Lewenstein from Cornell's Department of Science & Technology Studies. The network led by the CNF consisted of 12 sites and one affiliate. Since then it has grown to 14 sites. The CNF's original application included a mandate to "build the intellectual and institutional capacity needed to deal with social and ethical issues as they arise," emphasizing the need for situated studies and specifying that a number of social scientists would be embedded within the network so as to facilitate research and foment exchange and discussion (CNF 2003: 31). The CNF-led proposal eventually won the grant, and in the Spring of 2004, as I was completing my PhD, I interviewed for the position of Research Associate on the social and ethical issues in nanotechnology at the CNF/NNIN.

The job interview involved both STS and CNF/NNIN components. The STS part came first and focused on my research and my experience with ethnography. Bruce Lewenstein, Steve Hilgartner (both in the STS department), and I discussed the work that I would do at the CNF/NNIN and spent some time talking about the fantastic opportunity it provided to do ethnographic research on nano, as well as work alongside practitioners on new forms of collaboration. When we were done, Lewenstein escorted me to the newly inaugurated building where the CNF is housed. In contrast to STS's historical building, the CNF's Duffield Hall is new, airy, and stylish, with a high-tech and high-efficiency feel to it (Fig. 13.1). From experience, I know that speaking across disciplines is difficult and as I sat across from the lab manager in his cluttered office, watching him move his leg restlessly, I felt increasingly nervous. When, a few minutes later, the CNF/NNIN's Director[7] arrived, the interview started. Instead of a dialogue or intellectual exchange, we engaged in a period of "question and answer" that focused not on my research abilities but on what I would call 'instrumental issues'. For instance, one of the first questions was whether it was too early to study the social and ethical implications of nanotechnology, followed by a request to name these implications. I recall being taken aback by this line of questioning, which left me with the impression that it was not the Director's idea to examine these issues or hire someone like me. There was an implication that nano was both a stable construction (worthy of funding and research) *and* not an entity at all. Politely, I replied that I did not think this question

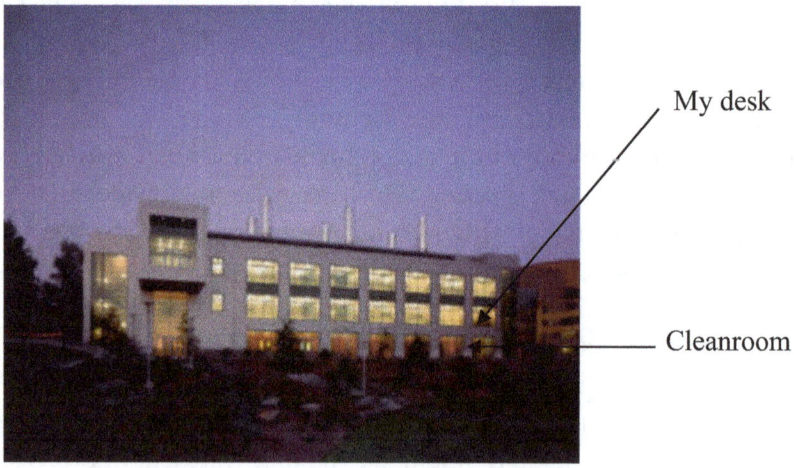

Fig. 13.1 My desk and the cleanroom at Cornell's Duffield Hall where the CNF is located

[7]At the time of my hiring, the CNF and the NNIN had the same Director. Later these duties were split.

could, or should, be answered without further studies of nano, its politics, technologies, funding, culture, et cetera. I then gave the standard STS answer that it is important not to speak of implications but think instead of dimensions, so as to highlight the co-production of the social and the technical. I left the CNF certain that I would not get the position. In December, I moved to Cornell.

To be sure, current efforts to integrate issues and concerns of the social sciences and humanities in technoscientific efforts make increased demands on scientists. I could no more expect my interviewers to possess a degree in STS than I could offer them one in nano-engineering. In this sense, the much maligned "deficit model" does work both ways. To state this, however, is to say little. This moment is significant because it points to a problematic reality-making strategy that would be employed for the next three years: a strict, hierarchical interaction with questions that require conventional, immediate, supposedly objective answers about what counts as a social, ethical, or scientific issue.

For the next three years, my job involved responding to and managing the distinct expectations these entities—the CNF, the NNIN, and my STS advisor—had of what caring for nanotechnology (and for the CNF/NNIN) meant. During this time, I attended the weekly staff meetings, went to staff lunches, celebrations, and picnics, took the CNF's obligatory safety course, led cleanroom visits, learned how to use (some of) the cleanroom equipment, helped organize events, and spoke at events organized by others. In 2007, I helped organize the CNF's 30th anniversary symposium and later conducted oral history interviews with numerous participants, including policymakers, social and natural scientists, and CNF staff members (Viseu and Maguire 2012). Throughout my tenure at the CNF, I learned much and took notes on my new knowledge and observations. The analysis that follows is based on these notes, documents, and experiences.

As I began my job, I thought my integrated status would afford me an opportunity to engage in new forms of collaboration with the CNF community that would be broader than the "ethics" model. I also thought that the STS and CNF communities both acknowledged and took advantage of the many ways in which science is already social and cultural. With some amount of naïveté, I thought this would be a "win-win" situation: I would learn about the practices of doing nano and the CNF's culture, and they would learn more about the social dimensions of their work; together we would produce collaborative practices and knowledge. While I knew that this was not going to be an easy task, I did not anticipate the rigidity of the definitions and expectations of the role that awaited me. The account I am thus able to produce is less about the making of nanotechnology than about the production of the social scientist as the silent and undervalued "care taker" of nanotechnology. My story is not new; it is one of enchantments and disenchantments, it is a story of everyday interactions that are laden with power asymmetries. It serves as a word of caution against interpreting integration-as-governance as a mode that privileges situatedness and collaboration, while overlooking the ways integration reifies boundaries and stereotypes. The intricacies of my three-year experience at the CNF/NNIN cannot be done justice here. However, I present four vignettes to illustrate some of the power-laden mundane interactions that defined my role in the facility and speak to the question of how to care for nanotechnology. They are exemplars and parables that elucidate the troubles of embeddedness alluded to above.

13.3.1 Being Integrated or, I Am Outreach

Having recently celebrated its 35th anniversary the CNF is, according to a staff member, the oldest end-to-end nanofabrication facility in the United States, standing as an exemplar of nano's legacy and age. The mission of the CNF, then Director George Malliaras told me, is to "proliferate the benefits of nanotechnology and make the facility available to a wide variety of people." In other words, the goal of CNF (and to a large degree that of the NNIN) is not to do research, but to provide the means for users to do so. To that end, the CNF possesses a cleanroom with nanofabrication equipment worth upwards of US $100 million, employs about 30 full time staff who provide users with assistance and expertise, and runs an operating budget of over US $5 million.

Shortly after arriving at Cornell University, I made the CNF[8] my home base and started to attend the facility's functions. At my first weekly staff meeting, I was introduced as the "ethics coordinator" and was told that I would be given a few minutes toward the end of the meeting to describe my work. As I scanned the room, I counted 25 bodies and gladly noted that six were women (myself included), which was, I thought, not a bad ratio for a technical enterprise. Halfway through the meeting, the lab manager announced, "ok, the administrative staff can leave" and with that four women stood up and headed toward the door. The women who stayed behind fulfilled "typical" female roles: I did "ethics", and my female colleague did "biology". The question of gender in science has long been a topic of exploration by feminists (Fox Keller 1987; Hubbard 1988) and here I found myself implicated in the embodied replication of traditional assumptions regarding gender, divisions of labor, and implicit hierarchical relations between the social and the natural sciences (Barry et al. 2008). The insistence on (wrongly) identifying my expertise and role in the facility as an ethicist added to my discomfort, as it was consistent with what Jasanoff (2011) characterizes as the "turn to ethics [as a] new mode of public reasoning" (p. 633). Ethics then further separates (technical) facts from (subjective) values, thus reinforcing expectations about what and how I would contribute; namely, research on the moral (and public) reasoning on the implications of the nanoscale work being done at the facility. I did not recognize it at the time, but together the subordinated, values-oriented gender division contributed to further defining ethical practice as the site of care work and cast me, an untenured

[8]As mentioned above, the CNF was then the organizing node of the NNIN. Unless specified, I will use the two interchangeably since in my daily experiences they were one and the same.

female social scientist, as the care-taker. This in turn added to the liminal status of my position and identity within the CNF. As the CNF Director told me,

> [The CNF] is an effort that is primarily… 95% of the budget is for infrastructure and running that infrastructure. And the ethics, at least within this side in the NNIN, is a small fraction of it. That's why there's only one of you and not 5 of you. There's only 1 Doug[9] and not 5 Dougs.

These a asymmetries are built into the structure of integration; they are not accidental. The bulk of the money and power remains with the technical staff. By implication, the marginal and subordinated role attributed to the social sciences is by design as well.

When it was my turn to speak, I briefly described my background and proceeded to explain that I was not an ethicist but a social scientist, and that my goal was to examine the situated practices through which nano is produced and enacted. I explained that this entailed studying the functioning of the CNF as an institution, as well as collaborating with the individuals who work in it. I stressed that I did not see my role as encompassing normative judgments of nano, nor was I interested in evaluating individual performances. Instead, I wanted to foster new kinds of scientific practices that made explicit room for their social, ethical, and cultural dimensions. No questions? The meeting was adjourned. I recall the conflict I felt when describing the work I wanted to do at the CNF, namely that of studying the workings of the facility. Managing relationships with subjects is never easy (Brettell 1993; Forsythe 2001), and here it was compounded by both the facts that I was on the facility's payroll and that, besides seeing subjects, I also saw future colleagues. Being paid to care complicated my relationship with the field site in ways novel to me, and I anticipated problems in separating identities, roles and importantly, affective entanglements.

As the weeks went by, I found my colleagues to be cordial yet distant, as if they did not know what to make of me. Since I wanted to be able to collaborate with them, as well as conduct ethnographic research, I decided to learn about the processes of nanofabrication. After taking the obligatory safety training course, I asked a colleague to take me into the lab and teach me how to nanofabricate. We spent three nights in the cleanroom working on two projects: nanocantilevers and carbon nanotube "trees". My notes are replete with explanations, schemes, and drawings. I recall one exchange as particularly telling: we were by the scanning electron microscope and I was having trouble understanding how it worked. My colleague touched the table we were leaning against and said, "this table, for instance, has an atomic structure like this" proceeding to draw it on a piece of paper. I remember realizing then that I was being introduced to a new language, and a new way of seeing. My colleague seemed to be experiencing something akin as he continuously probed me on my background, "A philosopher? A sociologist?" Our first night in the lab ended with me providing him an STS reading list at 3:30 in the morning.

---

[9]He is referring here to Doug Kysar (Law, Cornell) who substituted Bruce Lewenstein (Science and Technology Studies, (STS)) as my direct supervisor when the latter left the initiative in 2006.

The next day I was late to arrive at CNF and, as I walked in, I noticed that peoples' attitude towards me had changed. Everyone smiled (as they always did), but instead of simply nodding to acknowledge my presence, they would say, "I heard you did well in the lab yesterday!" or "I heard you wear your suit well!", complimenting me on my being careful in the lab. That was my first learning experience: I discovered that through the scientists' and engineers' gossip I had established some scientific credentials, making it easier for my colleagues to find common ground when relating to me. Later in the evening, when I joined my colleague in the cleanroom, he had some news. He told me that he had spoken to the CNF/NNIN's Director about my lab incursion, and that the Director was very excited about it and was going to make sure to include it in the CNF's yearly review report as an outreach activity. And that was my second lesson of the day: I discovered that I was outreach. I was part insider, paid staff with all the obligations that this entails, and part outsider—a sympathetic member of the public whose contribution to the facility was less one of expertise than one of willingness to be educated in the proper way to do and think about nanotechnology. Classifying me as "outreach" made me into a recognizable entity that could be managed and evaluated. In significant ways, I was a success story for integration: a measurable indicator, a compliant hybrid actor whose existence both fits and validates the requirements for integration as defined by Roco and Bainbridge (2001). This position was echoed in interviews that I later conducted with nanopractitioners who also saw me, and my existence, as outreach activities (Viseu and Maguire 2012).

13.3.2 Defining Integration or, 2 + 2 = 4

In the summer of 2004, Lewenstein and I were prompted by the CNF to participate in the NNIN's "research for undergraduates" program. We proposed a project that would use qualitative methods to produce a video exploring the "social and ethical issues" of nanofabrication at the CNF, and selected an engineering student to work with us. On the first day, as the undergrads trickled into the facility and started to meet one another, the following conversation took place outside my cubicle:

> Undergraduate: So, your mentor, is she a real scientist?
>
> My mentee: She is in the humanities…
>
> Undergraduate: Oh, so she doesn't know *anything* about all of this…
>
> (her emphasis)

Worried that it was the students who might not know much (or anything!) about the social sciences, and since our video was to be shown and discussed by all 81 undergraduates taking part in the nationwide program, Lewenstein and I emailed them all a reading package. The package included two readings: the first, by Lewenstein (2005), argued that what counts as a social issue in nano should be allowed to emerge from research instead of being defined upfront. The second, by

Vinck (2003), discussed the (fictitious) experience of a student's internship in a European laboratory, showing that doing science is not only about sitting at the lab bench, but also about politics and diplomacy.

The day after sending out the reading package the CNF/NNIN's Director approached me at my cubicle. Since this had never happened before, I was immediately on alert. "I did not like the text you sent the students and neither will they", he said. I asked why and he went on to discuss the dangers of theory. Then he concluded, "It's like in the old days in math class where they taught the language of math instead of teaching [students] that 2 plus 2 equals 4." I explained that one of the articles in particular presented students with the example of someone who was in a situation similar to theirs. In doing so, I said, it introduced them to less obvious (social, ethical, cultural, and political) dimensions and requirements of their work. The CNF Director looked unconvinced and replied that he would like to know what the students thought, and that he hoped "we had good survey results" (we were not planning to do a survey). I mumbled an "I'm sorry you feel that way", and with that he turned his back and left. I, in turn, ran anxiously to the phone to explain to Lewenstein what had just happened. Part of the Director's objection might have been based on pedagogy—rather than emphasize theory, he might have preferred an emphasis on empirical advice and examples—this is, after all, in line with his questions at the initial job interview. But my inability to provide a list of issues that would guide scientists in their "social and ethical issues" incursions, thus contributing to separating these issues from technoscientific practice, remained a sore point throughout my stay at the CNF, an implicit critique that helped define me as an underachiever.

In the aftermath of this episode, it became apparent to me that for the CNF/NNIN's Director, I was adding unnecessary and unwelcome complexity to nanotechnology, when I was expected to take care of that caring work myself. I worried that for him the value of integrating "the social and the ethical" did not include increasing understanding of the dynamics of lab research and innovation, with recognition of science as culture and its interactions with the wider society (see also, Doubleday and Viseu 2010). Instead, integration meant managing a narrow listing of possible risks and consequences, such that if a scientist followed the instructions and ticked off all the boxes, she was "social and ethical" and could go into the lab without having to worry again. To put it differently, the point was not to share the responsibility for the development of a caring science and technology, nor was it to use my knowledge to enrich current practices, but rather to find minimal, nondisruptive ways to sort through social and ethical issues. This was not a job that I could or wanted to do.

Rabinow and Bennett (2012) argue that the lack of open disagreement in integration engagements of social and natural science is problematic because it is through such disagreement that knowledge advances. They say that dissonance is a welcome shift from the indifference that characterized their own integration experience with synthetic biology. I share both their view and their experience. Yet in important ways, their argument evades the issue of power—dissonance is generative when actors are equals. In the example above, disagreement was not an opportunity to exchange ideas but a means to shut them off. It was also a display of power that I could not afford to ignore. This episode functioned both as a moment of recognition—an instance when the work I produced mattered enough for someone to notice it—and as a means to discipline me, ensuring that all definitions and asymmetries remained the same. This is not to say that I was powerless. Marginality, as feminist scholars have long recognized (Harding 1993; Hill Collins 1986), presents possibilities for subversion as well as a standpoint for knowledge creation. I made use of my marginal to draft an introduction to our website specifying that the social and ethical issues of nanotechnology are emergent and cannot be itemized a priori. This text remained online for years after I left the CNF/NNIN.[10]

13.3.3 Doing Integration or, the Problem with Users and Their Proxies

Users, I learned through the course of my stay, are both the CNF's raison d'être and a continual source of problems. The facility's users are an "in-between public"—more informed and interested in (nano)science than the public-at-large, yet lacking the knowledge and skill to afford them the insider/expert label. To ensure their safety, as well as that of staff, the cleanroom, and the environment, everyone who uses and works at the CNF must abide by the facilities' safety rules and protocols. A 68-page manual detailing them can be found online (Knight Laboratory 2013), and, most importantly, before going into the cleanroom, all new users attend a mandatory one-and-a-half day orientation and safety course, followed by a test. Yet, inevitably, violations to the rules take place and these are discussed at the weekly staff meetings. Not all rule violations are the same, but neither are all users: some—mostly those who have been around for a while and have demonstrated their technical and scientific proficiency—are respected as almost-colleagues. Others are seen as safety risks. The vast majority of users are known by name to the staff. Safety discussions are then both formal and personal; they are discussions of rules and of people.

During my stay at the CNF/NNIN, a decision was made to create training videos for some of the tools that populate the cleanroom, as a further mechanism to ensure proper usage and safety compliance, and prevent accidents. At one particular staff meeting, the question being debated was, "what is the best way to make these videos?" A lively discussion took place with talk of procedures to be highlighted, known problems with the tool, and common user difficulties. I listened to the conversation, and after a while, and since I was formally the facility's social and ethical coordinator, I spoke up to point out that *we* (I used the term "we"

[10]A modified version can be found here http://www.nnin.org/society-ethics/about-sei/intro-sei (accessed on 5 August 2015).

purposefully) were being remiss about not including a most important actor in our conversation: the users (Oudshoorn and Pinch 2005). In particular, I suggested that we consult with users on the difficulties they encounter with different machines. Giving them an active voice in the process—rather than making them visible only through third persons accounts of their problems—made all the more sense (to me) since they were also the target audience for the videos.

As soon as I was done talking, a silence and almost palpable reproach overtook the room. My colleagues looked at me with disbelief until someone finally broke the silence, saying that he was not sure we would want to do that—"there are *so* many problems with that," he said—and, without his needing to specify further, everyone else agreed. At this point, the facility's manager said that he was not sure how much we wanted to include users in the process, but that we could, perhaps, talk to those users that are "accommodating." With that, the conversation was over, leaving me as the lone, out-of-tune voice in the room.[11] Separate and apart from this, contacting those users who are "accommodating," that is, who have already been "mobilised" (Callon 1986), served the double purpose of appearing to take my intervention onboard while missing out on its main point—involving users who are experiencing issues with the equipment, and who may cause safety problems while doing so.

This speaks to built-in problems of the current integration design, namely its asymmetries in numbers and in power. Being the lone, postdoc female social scientist makes it hard or impossible to diversify and transform entrenched ideas, not only about science and its practice, but also about the social sciences and their possible contributions to the production of knowledge (see also Klenk and Meehan 2015). Emphasizing my knowledge and expertise implied furthering the distance between me and my colleagues, making it harder to justify my presence and usefulness to the facility. My identity and role within the CNF, never fully fitting and never independent, became, when actively exercised, even more unstable and precarious. This was especially so when, as the example above shows, my expertise disrupted existing beliefs. In other words, my presence was tolerated and even welcomed, as long as what I said and did remained within the confines of my own practice and did not disturb the facility's technoscientific culture. It was not only that my knowledge was undervalued—that certainly is not uncommon in interdisciplinary ventures—but that asserting it, especially when there was no-one to back me up, made my posture seem defensive and existence more precarious. Asserting my knowledge further drove a wedge between me and those who paid my salary, such that shame-inducing silence became my most attractive option.

[11]A similar thing happened when, months later, we were discussing the need to hire a student to help, and I suggested that given the low female–male ratio, we should make an attempt to hire a female since we had an excellent female candidate.

13.3.4 *Managing Integration or, What Counts as Knowledge*

In the summer of 2005, I was asked to speak at the "Short Course on Technology and Characterization at the Nanoscale" that the CNF organizes every year. The course is attended by staff as well as current and prospective users, giving me a good opportunity to address two pervasive assumptions I was encountering in my daily exchanges with colleagues and users, namely that "the social" and culture are not found inside the lab (and thus that my job, as interesting as it might be, remained outside of their expertise and practice), and that the social sciences' contribution to R&D is personal, normative, or moral. I decided to use the knowledge I was acquiring in my day-to-day observations as the basis for my arguments, thinking that this would help create a common ground and thus make it easier for those in attendance to relate to my talk. I carefully chose three topics that pointed to general issues rather than particular persons: controversies over definitions of contamination, negotiations over professional identity, and issues of tool ownership.

The talk seemed to be well received and was followed by a lively, if mostly off-topic, discussion on government funding, ethics and engineering, and why the public was so afraid of nanotechnology. It was thus all the more surprising when at the next staff meeting, the CNF/NNIN Director made some shrewd remarks about my undercover note taking. Announcing that I "was writing this [meeting] down," he mused aloud that he "guessed this was a public meeting" and that everyone had to be careful about what they said when I was in the room. Depicting me as the evaluator who sits in the corner, he evoked the suspicion that I was an outsider interested in judging what was being said and done, collectively or individually, and that I would later make my judgments public.[12] Since he was the CNF's main authority, and since, despite his obvious dislike he was not telling me to stop, he also made it sound as if he did not have the power to make me stop, thus reinforcing mistrust of my different, and potentially antagonistic, status. If, on one hand, this is not unusual (Latour and Woolgar 1986 [1979]) and could even be understood as an implicit authorization to continue conducting my ethnography, it also displayed a lack of institutional trust and support. It was as if looking inwards at the lab bench, at the very workings of the CNF, summoned me as an embedded adversary looking for troublesome rather than positive disruptions to existing practices.

---

[12]In what ways does this essay make public judgments, and how do I feel about this? Lucy Suchman, as special issue editor [of the journal where this article was originally published], asked me to reflect on this question. Part of the answer, I think, is covered by time, namely the time I waited before finding it 'safe' to write publicly about this, safe for both me (and my career) and for my CNF colleagues. Another part has to do with my essay's goal, which is not to do an exposé of a problematic site but to address the larger, systemic, endemic problems with the integration model. Still, of all that I have written, this text *still* raises the most questions for me: regarding my responsibility as an STS scholar and my responsibility toward subjects/colleagues, and regarding public evaluations of me and my work.

This adversarial attitude made itself visible again when, in the month prior to my departure for a job at a Canadian university, my request for a copy of the CNF/NNIN's original NSF proposal was rejected. (I later asked the overseeing NSF program officer for this proposal, and he too said "no".) One could be led to assume that the CNF was displeased with me, or my performance. But this is not the case. When, in the summer of 2007, I announced I would be leaving for a tenure-track job, the CNF's management asked me to stay onboard for another six months. While I cannot know with certainty, these inconsistent actions could perhaps be explained by a fear that when out of bounds I would be unmanageable. Equipped with three years of 'insider' knowledge, I would be a danger, a whistleblower who could harm the CNF/NNIN and nanotechnology.

These instances speak to my lack of power to define what counted as a proper site of care, and as such poses questions regarding the kinds of knowledge that can be produced within and through formal, institutionalized integration. Lab ethnographies, one of the "foundational pillars" of STS (Doing 2008), involve turning one's integrated gaze inwards, a move that, as I have shown, was received with fear and distrust. Separate and apart from this, ethnographies do not lend themselves to easy evaluation. They are open-ended, time consuming and slow to produce results. Even when concluded, they tend to offer few normative or instrumental answers. All this combines to make them poor evaluation metrics in a landscape ruled by the reigning logics of instrumentality and quick return on investment. Funding agencies expect results, often in the shape of "deliverables"; funded entities must oblige if they want their funding renewed, thus leaving little room for alternative or open-ended, qualitative research. Not surprisingly my contribution to the CNF/NNIN's output metrics was always carefully steered towards the creation of a web portal featuring the NNIN's activities and accomplishments around nanotechnology's "Social and Ethical Issues" (SEI). The portal, I was told, was to feature a nano-SEI reference database that was expected to be comprehensive and up-to-date. I understood the importance of having a web presence, but objected to the database. I argued that other research groups were exclusively dedicated to doing this and that we would, in effect, be duplicating their work with inferior quality. I quickly found out the database was nonnegotiable. As a public interface, the portal served as a measure of outreach; as a tangible infrastructure, it stood as a symbol of the CNF/NNIN's commitment to the responsible development of nanotechnology. It was always featured as the main item in the review reports written for the NSF's yearly evaluations. In other words, the portal offered itself as a deliverable that could be easily measured and quantified, such that the larger the database, the bigger the commitment. In so doing, it seemed to remove subjectivity from the "social and ethical", transforming it into a quantified, objective effort. In this translation, I came to impersonate the database, since I was the person charged with caring for it, and it came to impersonate me—with its size and health reflecting my performance and professional identity. The database translated my work and knowledge into a visible, understandable, and manageable deliverable, an entity that could be displayed in the review reports and transformed into metrics in ways that qualitative, open-ended, time-consuming, ethnographic research cannot.

13.4 Conclusion

As policymakers in the Western world increasingly embrace concepts of 'responsible research and innovation', integration goes mainstream. Science policy and funding guidelines have come to include calls for integration of the social sciences and humanities in technoscientific research and development projects to maximize societal benefits while minimizing negative impacts and public controversy. Articulating integration as both a kind of care work and as a 'matter of care' (Puig de la Bellacasa 2011) allowed me to evince the workings, premises, and goals of integration as well the associated and often invisible personal, epistemic, and affective costs of this world-making practice.

It could be argued that some of the issues I examine here are long-standing, and that access always comes at a price, often with "formal and informal provisions on what can be done and said" (Hackett and Rhoten 2011). Indeed, some of the details and conclusions drawn here are probably depressingly familiar to anyone who has ever done fieldwork—and particularly lab ethnography. Yet, as I have shown, there are important differences between the kinds of collaborations that can be fostered through traditional ethnographies, where the researcher has the ability to move between belonging to, and being removed from, the research site (Strathern 1999), and the ones that result from institutionalized calls for the integration of social scientists in R&D facilities, centers and projects. Being positioned within the initiative offers (in theory) the possibility of greater influence, but it also leaves the integrated researcher more vulnerable to pressure, less independent, and less able to produce and enact a science studies based knowledge and identity.

Some of the problems that I experienced are by design systemic and related to the "care" work I was hired to do. Integration works to eclipse difference, yet difference is ever present. It was present in the unsettling feelings and silences that occurred when the expertise I had to contribute was forced to be read as critical, adversarial, or ungrateful to the people with whom I worked and whose feelings and good opinion I cared about. It was present in my status as staff and outreach. It was present in my training as a qualitative scientist and my desire to do ethnographic research. But it was also silenced in everyday activities and yearly evaluation reports, such that we could speak in only one voice. Asymmetries are also built into integration (it is no coincidence that it is the social sciences who are the subject of the integration), but are made invisible in policy documents and STS reports. Asymmetries in funding, in numbers, and in power, together with an overall utilitarian orientation that focuses on "deliverables", consensus, and quantifiable results, work to constrain the ability of integrated social sciences to define and implement goals, questions, methods, or findings. I note that my condition as a young, untenured woman exacerbated the power asymmetries at work. However, far from making my tale less generalizable or less worrisome, thinking with care exposes the need for increased scrutiny of the dynamics at play so as to enhance the effectiveness and livability of this increasingly common role. Within this framework, caring for nanotechnology is another way to reinforce a divide between facts and values. In other words, despite integration's potential for creating new forms of collaboration between the social and natural sciences, its discourses and socio-material orderings are based on traditional and prescriptive arrangements, where disciplinary boundaries, funding arrangements and power asymmetries are not challenged but reified such that there is little to no room to re-imagine existing practices. Integration in its current format is problematic and must be reassessed.

Failure and success have, in STS circles, become relative. As Latour's *Aramis* (1996) shows us, both are in the eye of the beholder. The same can be said of my experience of integration. Certainly, there were successes—some of them described above—and, in some way, the very publication of this chapter is a measure of success (see also, Rabinow and Bennett 2012). Yet, as integration and the in-house social scientist model gain traction, we must question their goals and implementation. If our goal is to work towards the opening of science to other actors, questions, and methods, to foster new collaboration models, and to facilitate shared knowledge, then the institutionalized integration of social scientists within R&D initiatives is designed to fail because in its current enactment, social scientists are being called upon to care for nanotechnology by keeping it undisturbed, and doing care differently is understood as a threat. Moreover, when failure is by design personal failure—entangled with one's identity and career trajectory—the affective cost of integration is exceptionally high.

The road ahead is as difficult as it is uncertain. We must continue working toward fostering scientific practices that are inclusive of commitments to social and ethical issues. Doing so requires a continued effort to transform current integration policy so as to include and facilitate broader, less utilitarian enactments of the scientific work, social sciences, and (care) work they are slotted to perform. Some "easy" fixes may be insisted upon: integration should accommodate teams of social scientists, rather than one or two individuals;[13] funding for these teams should not be controlled by, or depend on, the site of integration; the teams must be given the power to define and implement their activities; and, last but not least, evaluation of their activities must be conducted by peers. This is a starting place, yet there is more work to be done. We must insist on the value of complexity, such that results from these collaborations do not need to be presented in one voice, or in one metric-friendly deliverable. We must continue to move away from a framework where collaboration with social scientists is adversarial rather than positive, and where success is equated with consensus. We must continue to work toward an understanding of science that is inclusive of affective, response-able (Haraway 2007), and shared relationships, such that dualisms are replaced by entanglements.

As we work toward this we in the field of STS must also take a closer look at the positions we adopt toward and within the worlds we study and co-construct. Positionality has long been recognized by feminist science scholars such as Hill Collins (1986), Haraway (1988), and Harding (1993) as key to epistemology and knowledge production because all knowledge is locally situated. Positionality is

[13]I am reminded here of Wynne's account of withdrawing from a government program because "acting without STS allies [he] was utterly unable to diversify existing entrenched ideas about innovation and future expectations" (Wynne 2007: 497).

also intimately connected to power and knowledge such that some entities and groups remain central while others are marginalized and neglected, thus limiting or enabling the production of particular kinds of knowledge (Hill Collins 1986; Harding 1993). Reflecting on her years working as an anthropologist at Xerox PARC, Suchman (2013) argues that where researchers are located "is perhaps *the* critical question" (p. 147) as it speaks to issues of sponsorship (who pays for the research) and accountability (who is responsible for defining the research questions, methodology, conceptual framework, et cetera) that articulate and afford different frames of reference and values. It is naïve, she suggests, to expect the frames of reference of all actors to be the same. As she goes on to illustrate, in the process of negotiation, differences are often obliterated in the interest of speaking in one voice. Thus, she says, in an ironic twist of fate, the more successful a project is, the less room there is for configuration and difference, and the more likely it is that STS knowledge is appropriated (Suchman 2013; see also Wynne 2007; Joly and Kaufmann 2008). Likewise, Rabinow and Bennett (2012) end their account by recommending to others that they find a position that allows them to "remain adjacent" (p. 177) to the initiatives they study. Reflecting on positionality involves, at the very least, speaking aloud of the troubles of integration: the kinds of traditional and prescriptive arrangements that frame and gauge its success, the instrumentalization of social science work it imposes, the differences it eclipses, and its existential, epistemic, and affective costs. In a world of decreased funding for social sciences and humanities, speaking out of tune is both difficult and crucial. It may also involve a return to a position outside of the spotlight of funding bodies and policy agencies.

Ideally, we would see 'stand-alone' funding for the social sciences increased without requirements for integration or subordination to a given topic or big science initiative. But in the current context this seems unlikely. In its absence, and in a position that is likely to be unpopular among STS scholars, I argue that we should push to a return to funding structures that resemble those of the HGP's ELSI program. The ELSI program was part of the HGP (and its funding was subordinated to the HGP's budget) but worked as a granting entity with its own administrative and funding structures. I am not advocating replicating the ELSI model, with its emphasis on after-the-fact impacts, and its enduring (and warranted) lack of independence that precluded questioning of the very project that set it up. Yet, I would like to point out that despite these (and other) handicaps ELSI was key in funding numerous important pieces of critical scholarship (e.g. Kay 2000; Sloan 2000; see also, Lindee 1994). I am not sure the same can be said of nanotechnology, where criticism of funding arrangements, for instance, has been remarkably absent. I am arguing that we *build upon* (not replicate) the ELSI model's organizational structure and work to ensure that we create granting bodies that have some degree of autonomy and are adequately staffed, funded, and importantly, evaluated. Nonengagement with the world and its subjects is not an option. But we must remain analytically skeptical and find positions where collaboration and the situated knowledge generated within those collaborations is a starting point for critical questions and reassessments.

13.5 Afterword

I end this chapter with an attempt to answer the question that I am invariably asked when presenting the material upon which this chapter is based: given the current integration policy framework, what is the one thing I recommend we do to succeed in our engagements and collaborations?

In the three years since the publication of this article we have witnessed two concurrent and paradoxical developments: on one hand, integration, consolidated into the label of "responsible research and innovation" (RRI), has become firmly rooted as a mainstay of technoscientific policy (see, for instance, de Saille 2015a; Fisher forthcoming; Horizon 2020 nd). On the other hand, there has been an increase in the number of articles that examine the difficult workings of integration as governance (see, for instance, Giordano 2018; Klenk and Meehan 2015; de Saille 2015b; Ribeiro et al. 2017).

Many of the critiques currently being leveled at integration are in line with my argument and past experience, and I will not repeat them here. Recently, however, integration has been studied in its connections to distinct political models and technologies. This is an important argument that deserves attention. In their article, Klenk and Meehan (2015) argue that integration is foremost an "idealization of how scientists 'should work' with the diversity of actors/stakeholders" with the goal of producing "normative knowledge and policy-relevant solutions aimed at societal problems" (p.160). If this formulation sounds familiar it is because it replicates the discourse of Western policymakers and funding bodies. For instance, Horizon 2020—the European Union's current research and innovation funding programme—describes the rationale for RRI stating,

> Responsible Research and Innovation (RRI) implies that societal actors (researchers, citizens, policy makers, business, third sector organisations, etc.) work together during the whole research and innovation process in order to better align both the process and its outcomes with the values, needs and expectations of society. (H2020 n.d)

In other words, in the face of a complex and heterogeneous world, integration is figured as facilitating (instrumental) technoscientific solutions directed at societal problems. Klenk and Meehan (2015) go on to argue that if integration is to blur disciplinary lines then it must address the politics of knowledge. Yet, as I have argued, integration tends to hide the messiness of dissenting voices and of alternative ways of knowing, favoring instead the consensus (see also, Viseu 2015b). Klenk and Meehan (2015) conclude that "the integration imperative conceals the friction, antagonism and power exclusion inherent in knowledge co-production" (p. 161). I would argue that not only do the politics of knowledge production remain invisible—above I described that integration tends to hide the messiness of dissenting voices and alternative ways of knowing, favoring instead consensus—what is increasingly evident is that integration is by design blind to politics and power. Moreover, as argued by Giordano (2018), "the superficial language of rights and democracy [used in current RRI policy documents] relegitimizes the primacy of scientific discovery to solve societal problems" (p. 401). Thus, in ironic but not unexpected ways, integration (and the corollary "responsible research and innovation") may work to place more, not less, authority in a few, expert hands while simultaneously reinforcing the divide between the social and the scientific expert knowledge.

In his book, Laurent (2017) examines the links between nanotechnology and democracy. Taking problematization to mean "the range of ways to tackle a problem" (p.18), Laurent argues that the ways in which nanotechnology is made into a problem speak to how democracy itself is problematized. His comparative analysis shows, for instance, how distinct instruments work to create informed and participating publics and/or deliberating citizens, in Europe and the United States. Importantly, Laurent describes the ways in which French experiments with nanotechnology worked to exclude the most active and participating public: the anti-nano activists. He is not alone in this conclusion. Both Giordano (2018) and de Saille (2015a, 2015b) caution that RRI may, in fact, be working to exclude unruly and critical publics.

Deciding how an issue is framed, what is privileged, who gets to participate and how, who (and what) is excluded—is at the heart of politics. Creating room for disagreement and alternative ways of caring and knowing is essential for and in a democracy. Yet, my experience, as described above, is that in its current enactments integration does not facilitate disagreement and may in fact hinder it (see also, Giordiano 2018; de Saille 2015b; Ribeiro et al. 2017). To facilitate collaboration and engagement in technoscientific development and innovation it is not enough to integrate so as to be representative, it is also necessary to attend to configurations of power, to funding and evaluation mechanisms, and vitally, to the generative role of dissent.

It is with the generative value of dissent that I end my story. When reflecting upon their integration experience, Rabinow and Bennett (2012) advise other social scientists engaged in integration experiments to "speak the truth frankly" (p. 179); Fitzgerald et al. (2014) provide the contrary advice: they suggest that collaborators make use of "equivocal speech", which they describe as an approach that is "attentive to the things that are better left unsaid, to the feelings that are as well off not articulated and to the senses of awkwardness and ignorance that probably will not help anything if openly acknowledged." (p. 716). I would like to suggest a third alternative: that we practice a politics of "collaborative dissent", that is, a mode of working-together that acknowledges distinct motivations and goals; that is welcoming of complexity and does not attempt to speak in unison; that does not equate a plurality of views with failure, but with achievement; that knows that dissent can be generative and enriching—and in fact, crucial—to a democratic society. Insisting on dissent and complexity will not eclipse power differences, but it may ameliorate them enough to facilitate the creation of distinct ways to care and produce knowledges.

Acknowledgements I would like to acknowledge support from the National Nanotechnology Infrastructure Network, where I worked from 2004 to 2007. I would also like to thank the anonymous reviewers who were generous, enthusiastic, and invaluable in helping me develop my thinking.

Funding This work was supported by the National Science Foundation under grant no. ECS-0335765.

References

Aguiton, S. (2012). Etre embarquée das une competition de biologie synthétique. Paper delivered at the 'Sciences Sociales Embarquées' Colloque International, organized by CETCOPRA (Université Paris 1) & CSI (Mines ParisTech), Paris, France, January 13 & 14, 2012.

Armstrong, P., & Armstrong, H. (2002) Thinking it through: Women, work and caring in the New Millenium. *Canadian Woman Studies/Les Cahiers De La Femme, 21/22*(4/2), 44–50.

Balmer, A., Bulpin, K., Calvert, J., Kearnes, M., Mackenzie, A., Marris, C., et al. (2012). *Towards a manifesto for experimental collaborations between social and natural scientists*. Available at: http://experimentalcollaborations.wordpress.com. Accessed November 29, 2012.

Barry, A., Born, G., & Weszkalnys, G. (2008). Logics of interdisciplinarity. *Economy and Society, 31*(1), 20–29.

Brettell, C. B. (Ed.). (1993). *When they read what we write: The politics of ethnography*. Westport, Connecticut & London: Bergin & Garvey.

Callon, M. (1986). Some elements of a sociology of translation: Domestication of the scallops and the fishermen of St. Brieuc Bay. In J. Law (Ed.), *Power, action and belief: A new sociology of knowledge? Sociological Review Monograph* (Vol. 32, pp. 196–233). London, UK: Routledge & Kegan Paul.

Calvert, J., & Martin, P. (2009). The role of social scientists in synthetic biology. *Science and Society Series on Convergence Research. EMBO (European Molecular Biology Organization) Reports, 10*(13), 201–204.

Cornell Nanofabrication Facility (CNF) (2003) National Nanotechnology Infrastructure Network Proposal: Submitted to NSF in response to NSF 03–519.

de Saille, S. (2015a). Innovating innovation policy: the emergence of 'Responsible Research and Innovation'. *Journal of Responsible Innovation*, *2*(2), 152–168.

de Saille, S. (2015b). Dis-inviting the Unruly Public. *Science as Culture*, *24*(1): 99–107.

Doing, P. (2008). Give me a laboratory and I will raise a discipline: the past, present, and future politics of laboratory studies in STS. In E. J. Hackett, O. Amsterdamska, M. Lynch & J. Wajcman (Eds.), *The Handbook of Science and Technology Studies*. Cambridge, MA: MIT Press, 279–298.

Doubleday, R., & Viseu, A. (2010). Questioning interdisciplinarity: What roles for laboratory based social science? In K. Kjolberg & F. Wickson (Eds.), *Nano meets macro: Social perspectives on nano sciences and technologies* (pp. 51–75). New Jersey: Pan Stanford Publishing.

European Commission. (2004). *Towards a European strategy on nanotechnology*. Brussels: European Commission. Available at: ftp://ftp.cordis.europa.eu/pub/nanotechnology/docs/nano_com_en.pdf. Accessed March 30, 2009.

European Commission. (2014). *Horizon 2020 in Brief: the EU Framework Programme for Research & Innovation. Luxembourg: Publications Office of the European Union*. Available at: http://ec.europa.eu/programmes/horizon2020/en/news/horizon-2020-brief-eu-frameworkprogramme-research-innovation (accessed 12 March 2015).

Fisher, E. (2005). Lessons learned from Ethical, Legal and Social Implications program (ELSI): Planning societal implications research for the national nanotechnology program. *Technology in Society, 27,* 321–328.

Fisher, E. (forthcoming). "Enhancing Micro-foundations of Responsible Innovation: Integration of Social Sciences and Humanities with Research and Innovation Practices." In: R Von Schomberg (ed) *Handbook of Responsible Innovation*. Cheltenham, UK: Edward Elgar.

Fitzgerald, D., Littlefield, M., Knudsen, KJ., Tonks, J. & Dietz, MJ. (2014). Ambivalence, equivocation and the politics of experimental knowledge: A transdisciplinary neuroscience encounter. *Social Studies of Science*, *44*(5): 701–721.

Forsythe, D. (2001). *Studying those who study us: An anthropologist in the world of artificial intelligence. Writing Science Series*. Stanford, CA: Stanford University Press.

Fox, B. (Ed.). (1980). *Hidden in the household: Women's domestic labour under capitalism*. Toronto: The Women's Press.

Fox Keller, E. (1987). The gender/science system: Or, is sex to gender as nature is to science? *Hypatia, 2*(3), 37–49 (Special Edition on 'Feminism & Science 1').

Gannon, F. (2009). Convergence (Editorial). *EMBO (European Molecular Biology Organization) Reports, 10*(2), 103.

Gibbons, M. (1999). Science's new social contract with society. *Nature, 402,* C81–C84.

Gibbons, M., Limoges, C., Nowotny, H., Schwartzman, S., Scott, P., & Trow, M. (1994). *The new production of knowledge: The dynamics of science and research in contemporary societies*. London: Sage.

Gorman, M. (2011). Doing science, technology and society in the National Science Foundation. *Science and Engineering Ethics, 17*(4), 839–849.

Guston, D. (2010, May). *Societal dimensions research in the national nanotechnology initiative* (CSPO Report #10-02). Arizona State University. Consortium for Science, Policy & Outcomes. [Online]. Available at: http://www.cspo.org/library/title/?action=getfile&file=291&section=lib. Accessed November 17, 2010.

Hackett, E. J., & Rhoten, D. R. (2011). Engaged, embedded, enjoined: Science and technology studies in the National Science Foundation. *Science and Engineering Ethics, 17*(4), 823–838.

Haraway, D. (1988). Situated knowledges: The science question in feminism and the privilege of partial perspective. *Feminist Studies, 14*(3), 575–599.

Haraway, D. (2007) *When Species Meet*. Minneapolis, MN: University of Minnesota Press.

Haraway, D. (2010). When species meet: Staying with the trouble. *Environment and Planning D: Society and Space, 28*(1), 53–55.

Harding, S. (1993). Rethinking standpoint epistemology: What is "Strong Objectivity"? In L. Alcoff & E. Potter (Eds.), *Feminist epistemologies* (pp. 49–82). London: Routledge.

Hill Collins, P. (1986). Learning from the outsider within: The sociological significance of black feminist thought. *Social Problems, 33*(6), S14–S32.

Horizon (2020). (nd). Responsible Research and Innovation. Available at: https://ec.europa.eu/programmes/horizon2020/en/h2020-section/responsible-research-innovation (accessed 12 June 2018)

Hubbard, R. (1988, Spring). Science, facts and feminism. *Hypatia, 3*(1), 5–17 (Special Edition on 'Feminism and Science 2').

Jasanoff, S. (2005). *Designs on nature: Science and democracy in Europe and the United States*. Princeton: Princeton University Press.

Jasanoff, S. (2011). Constitutional moments in governing science and technology. *Science and Engineering Ethics, 17*(4), 620–638.

Joly, P.-B., & Kaufmann, A. (2008). Lost in translation? The need for 'Upstream Engagement' with nanotechnology on trial. *Science as Culture, 17*(3), 1–23.

Karinen, R., & Guston, D. H. (2010). Toward anticipatory governance: The experience with nanotechnology. In M. Kaiser, M. Kurath, S. Maasen, & C. Rehmann-Sutter (Eds.), *Governing future technologies: Nanotechnology and the rise of an assessment regime* (pp. 217–232). The Netherlands: Springer.

Kay, LE. (2000). *Who Wrote the Book of Life? A History of the Genetic Code*. Stanford, CA: Stanford University Press.

Klenk, N. & Meehan, K. (2015). Climate change and transdisciplinary science: Problematizing the integration imperative. *Environmental Science & Policy*, 54: 160–167.

Knight Laboratory. (2013). *Cornell NanoScale Science and Technology Facility (CNF) Laboratory usage & safety manual*. Available at: http://www.cnf.cornell.edu/doc/CNF_Lab_Manual_10th_edition.pdf. Accessed April 10, 2013.

Latour, B. (1996). *Aramis or the love of technology*. Cambridge, MA: Harvard University Press.

Latour, B. (2004). Why has critique run out of steam? From matters of fact to matters of concern. *Journal of Critical Inquiry, 30*(2), 225–248.

Latour, B., & Woolgar, S. 1986 [1979]. *Laboratory life: The social construction of scientific facts*. Beverly Hills, London: Sage.

Laurent, B. (2017). Democratic experiments: problematizing nanotechnology and democracy in Europe and the United States. Cambridge, MA: MIT Press.

Lewenstein, B. (2005). What counts as a 'Social and Ethical Issue in Nanotechnology?' *HYLE—International Journal for Philosophy of Chemistry, 11*(1), 5–18 (Special edition on 'Nanotech Challenges' Part 2. Jointly published with *Techne*).

Lindee, S. (1994). The ELSI hypothesis. *Isis 85*(2): 293–296.

Lynch, M. (2000). Against reflexivity as an academic virtue and source of privileged knowledge. *Theory, Culture & Society, 17,* 26–54.

McCain, L. (2002). Informing technology policy decisions: The US human genome project's ethical, legal, and social implications programs as a critical case. *Technology in Society, 24,* 111–132.

Mol, A., Moser, I., & Pols, J. (Eds.). (2010). *Care in practice: On tinkering in clinics, homes and farms*. Bielefeld: Verlag.

Murphy, M. (2015). Unsettling care: Troubling transnational itineraries of affect in feminist health practices. *Social Studies of Science*, *45*(5): 717–737.

National Nanotechnology Initiative (NNI) (n.d). *NNI vision, goals and objectives.* Available at: http://www.nano.gov/about-nni/what/vision-goals. Accessed November 10, 2005.

Nordmann, A. (2007). Knots and strands: An argument for productive disillusionment. *Journal of Medicine and Philosophy, 32*(3), 217–236.

Nowotny, H., Scott, P., & Gibbons, M. (2001). *Re-thinking Science: Knowledge and the public in an age of uncertainty*. Cambridge: Polity Press.

National Science Foundation (NSF). (2003). *National Nanotechnology Infrastructure Network (NNIN) program solicitation* (Program Solicitation NSF 03–519). Available at: http://www.nsf.gov/pubs/2003/nsf03519/nsf03519.pdf. Accessed June 18, 2006.

Oudshoorn, N., & Pinch, T. (Eds.). (2005). *How users matter: The co-construction of users and technology*. Cambridge, MA: MIT Press.

Public Law 108-153. (2003). *21st century nanotechnology research and development act*. 108th Congress. Available at: http://frwebgate.access.gpo.gov/cgi-bin/getdoc.cgi?dbname=108_cong_public_laws&docid=f:publ153.108.pdf. Accessed June 12, 2008.

Puig de la Bellacasa, M. (2011). Matters of care in technoscience: Assembling neglected things. *Social Studies of Science 41*(1): 85–106.

Rabinow, P., & Bennett, G. (2012). *Designing human practices: An experiment with synthetic biology*. Chicago, IL: The University of Chicago Press.

Ribeiro BE, Smith RDJ and Millar K (2017). A Mobilising Concept? Unpacking Academic Representations of Responsible Research and Innovation. *Science and Engineering Ethics*, *23* (1): 81–103.

Rip, A. (2006). Folk theories of nanotechnologists. *Science as Culture, 15*(4), 349–365.

Rip, A. (2009). Futures of ELSA. *Science & Society Series on Convergence Research. EMBO (European Molecular Biology Organization) Reports, 10*(7), 666–670.

Roco, M., & Bainbridge, W. S (Eds.). (2001) *Societal implications of nanoscience and nanotechnology*, Final Report, National Science Foundation Workshop, September 28–29, 2000. Arlington, VA: NSF. Available at: http://www.wtec.org/loyola/nano/NSET.Societal.Implications/nanosi.pdf. Accessed February 07, 2004.

Schuurbiers, D. (2011). What happens in the lab does not stay in the lab: Applying midstream modulation to enhance critical reflection in the laboratory. *Science and Engineering Ethics, 17* (4), 769–788.

Schuubiers, D., & Fisher, E. (2009). Lab-scale intervention. *Science & Society Series on Convergence Research. EMBO (European Molecular Biology Organization) Reports, 10*(5), 424–427.

Shapira, P., Youtie, J., & Porter, A. L. (2010). The emergence of social science research on nanotechnology. *Scientometrics, 85*(2), 595–611.

Sismondo, S. (2008). Science and technology studies and an engaged program. In E. Hackett, O. Amsterdamska, M. Lynch, & J. Wajcman (Eds.), *The handbook of science and technology studies* (pp. 13–31). Cambridge, MA: MIT Press.

Sloan, PR. (ed.) (2000). *Controlling Our Destinies: Historical, Philosophical, Ethical, and Theological Perspectives on the Human Genome Project*. Notre Dame, IN: University of Notre Dame Press.

Stegmaier, P. (2009). The rock 'n' roll of knowledge co-production. *Science & Society Series on Convergence Research. EMBO (European Molecular Biology Organization) Reports, 10*(2), 114–119.

Strathern, M. (1999). *Property, substance and effect. Anthropological essays on persons and things*. London: Athlone Press.

Suchman, L. (2013). Consuming anthropology. In A. Barry & G. Born (Eds.), *Interdisciplinarity: Reconfigurations of the social and natural* sciences (pp. 141–160). London: Routledge.

T Kulve, H., & Rip, A. (2011). Constructing productive engagement: Pre-engagement tools for emerging technologies. *Science and Engineering Ethics, 17*(4), 699–714.

Thoreau, F. (2012). Being enrolled and being engaged back. Paper delivered at the 'Sciences Sociales Embarquées' Colloque International, organized by CETCOPRA (Université Paris 1) & CSI (Mines ParisTech), Paris, France, January 13 & 14, 2012.

Van Maanen, J. (1988). *Tales of the field: On writing ethnography*. Chicago and London: The University of Chicago Press.

Vinck, D. (Ed.). (2003). *Everyday engineering: An ethnography of design and innovation*. Cambridge, MA: MIT Press.

Viseu, A. (2012). Integrating the social: Being a social scientist in a nanotechnology laboratory. Paper delivered at the 'Sciences Sociales Embarquées' Colloque International, organized by CETCOPRA (Université Paris 1) & CSI (Mines ParisTech), Paris, France, January 13 & 14, 2012.

Viseu, A., & Maguire, H. (2012). Integrating and enacting 'social and ethical issues' in nanotechnology practices. *NanoEthics, 6,* 195–209.

Viseu, A. (2015a). Caring for nanotechnology? Being an integrated social scientist. Special Issue on 'The Politics of Care in Technoscience'. *Social Studies of Science, 45*(5): 642–664.

Viseu, A. (2015b). Integration of social science into research is crucial (World View). *Nature, 525* (7569): 291.

Webster, A. (2007). Crossing boundaries social science in the policy room. *Science, Technology and Human Values, 32*(4), 458–479.

Wolfe, A. (2000). Federal policy making for biotechnology, executive branch, ELSI. In T. H. Murray & M. J. Mehlman (Eds.), *Encyclopedia of ethical, legal and policy issues in biotechnology* (Vol. 1, pp. 234–240). New York: Wiley.

Wynne, B. (2007). Dazzled by the mirage of influence? STS–SSK in multivalent registers of relevance. *Science, Technology and Human Values, 32*(4), 491–503.

Chapter 14
The Art of Research: A Divergent/Convergent Thinking Framework and Opportunities for Science-Based Approaches

Glory E. Aviña, Christian D. Schunn, Austin R. Silva, Travis L. Bauer, George W. Crabtree, Curtis M. Johnson, Toluwalogo Odumosu, S. Thomas Picraux, R. Keith Sawyer, Richard P. Schneider, Rickson Sun, Gregory J. Feist, Venkatesh Narayanamurti and Jeffrey Y. Tsao

14.1 Introduction

Research, the production of new and useful knowledge and products, is an estimated \$1.6T/year world enterprise (Grueber and Studt 2014), supporting a community of approximately 11 million active researchers (Haak 2014), and, most importantly, fuelling a large fraction of wealth creation in our modern economy (Jones 1995). As a consequence, it is an essential function of research and engineering management, with a growing literature associated with how it can best be organized and facilitated (Damanpour 1991; Damanpour and Gopalakrishnan 2001;

G. E. Aviña · A. R. Silva · T. L. Bauer · C. M. Johnson · J. Y. Tsao (✉)
Sandia National Laboratories, Albuquerque, NM 87185, USA
e-mail: jytsao@sandia.gov

T. Odumosu
University of Virginia, Charlottesville, VA 22903, USA

G. W. Crabtree
Argonne National Laboratory, Argonne, IL 60439, USA

S. T. Picraux
Los Alamos National Laboratory, Los Alamos, NM 87545, USA

R. K. Sawyer
University of North Carolina, Chapel Hill, NC 27599, USA

C. D. Schunn
University of Pittsburgh, Pittsburgh, PA 15260, USA

E. Subrahmanian et al. (eds.), *Engineering a Better Future*,
https://doi.org/10.1007/978-3-319-91134-2_14

Drucker 2011; Hueske et al. 2015). From our perspective 'in the trenches' of research in perhaps the most mature and quantitative of the sciences, the physical sciences and engineering, research is still largely practiced as an 'art' where it passed down from the experience of one generation to the next and, as decribed in Beveridge (1957), focuses on the intuitive nature of the scientist. We learn how to do research from our professors, managers, mentors and fellow researchers, just as they did from theirs. To be sure, this art is highly evolved, and successful researchers and research managers have developed a deep understanding of how best to practice this art, won from hard-earned experience. However, this understanding is intuitive, without the reductionist framework necessary to analyse and decode best practices, improve and replicate such practices and then systematically raise the bar everywhere else.

Paralleling this community of research practice, another community has been growing around a field that might broadly be called the 'science' of research (Börner et al. 2010; Fealing 2011; Feist 2008; Sawyer 2011; Sismondo 2011)—the understanding of the human and intellectual processes associated with research. Until now, the two communities (the practitioners or 'artists' of research and the 'scientists' of research) have advanced with minimal interaction. In principle, however, they can benefit each other enormously. Practitioners of research care deeply about how effective they are, and what better way to improve their effectiveness than to apply scientific principles; while scientists of research care deeply about their scientific understanding of research, and what better way to test that understanding than to try to apply it to improving how research is actually done.

Indeed, in the larger domain of research policy, the National Science Foundation's Science of Science and Innovation Policy (SciSIP) programme aims to bring together scientists of science *policy* with artists of science policy (Fealing 2011). But that intersection leaves much on the table. Harnessing the science of science and innovation to improve not just science and innovation policy (the top-down organization of the research enterprise) but the research *itself* (the bottom-up practice of research) is lower hanging fruit, and more amenable to shorter cycle times on experimentation and learning and hence to faster progress. We believe the opportunity to discover and apply the principles governing the effective practice of research is large, and we call here for policy and funding support for such experimentation and learning. In fact, one piece of the intersection

---

G. J. Feist
San Jose State University, San Jose, CA 95192, USA

V. Narayanamurti
Harvard University, Cambridge, MA 02138, USA

R. Sun
IDEO, Palo Alto, CA 94301, USA

R. P. Schneider
glo-USA, Sunnyvale, CA 94089, USA

between the bottom-up practice of research and its scientific study is already being aggressively explored in the life sciences, through the National Institutes of Health's Science of Team Science (SciTS) programme (Börner et al. 2010).

We believe a similar opportunity exists in engineering and physical sciences (EPS), and that the physical sciences and engineering bring unique advantages both as an object of study and as an object of self-improvement. Physical science is the 'exemplar' science, arguably the deepest and most advanced, hence we have much to learn from how its research practice enabled it to become so. EPS is the most aggressively mathematized, reductionist and data-driven area of research, a perspective which, if reflexively applied to itself, could be uniquely productive. Though direct usefulness to society is not always the goal of research, when it is, the path to such usefulness is cleaner in EPS, e.g. not confounded by the regulatory processes associated with the life sciences. As the most mature of the sciences, at this point in time, EPS spans a huge range of small to large scale and of disciplinary to massively interdisciplinary, research, and thus would provide a severe test of any proposed framework for understanding and improving research. Finally, the physical sciences are hardly over with—many of our most pressing planetary-scale problems—from moving the world towards a sustainable energy diet to interconnecting the world's people and resources—require solutions rooted in the physical sciences and engineering. We argue for an urgency associated with improving how we do EPS research in service of these problems.

As our own initial step in this direction Sandia National Laboratories hosted a Forum and Roundtable, which brought together distinguished EPS practitioners of the art of research and experts in the emerging social science of research. The two communities engaged in a broad-based discussion, and concluded that there are indeed opportunities for reciprocal benefit. In this chapter, we build on that discussion and outline some of those opportunities. Some of the opportunities overlap with those already identified in the science and engineering management communities. However, since the opportunities emerged from these two different and newly interacting communities, some will be new to the traditional science and engineering management communities. Perhaps most importantly, as we organized new opportunities to collaborate between the two communities, divergent/convergent thinking emerged as a critical framework from the Forum and Roundtable.

14.2 Divergent/Convergent Thinking Framework

The framework rests on three overarching foundational assumptions, or hypotheses, that emerged from our Forum and Roundtable.

First: interactive divergent (idea generation) and convergent (idea test and selection) thinking are the fundamental processes underlying research (Cropley 2006). Here, we think of iterative and closely interacting cycles of idea generation followed by idea filtering, refining and retention (Toulmin 1961). Note there is no strict process ordering from divergent to convergent. Divergent and convergent thinking continue to apply as early theories or plans become increasingly more detailed and elaborate because details can be found to be problematic and in need of replacement with alternatives. Further, some gaps only become apparent as theories are elaborated or applied to new situations. For simplicity, we use for these complementary cycles the common terms divergent and convergent thinking, with the understanding that they are related (but not identical) to other terms used in the cognitive, social and computational sciences: idea generation/test, blind variation and selective retention (BVSR) (Campbell 1960; Simonton 2013), abductive versus deductive reasoning, generative versus analytic thinking, discovery versus hypothesis-driven science (Medawar 1963), creativity versus intelligence, thinking fast versus thinking slow (Kahneman 2011), foraging versus sense-making (Pirolli and Card 2005), exploration versus exploitation (March 1991) and learning/growth versus performance/fixed mindsets (Dweck and Leggett 1988).

Second: the quality, quantity and interactivity of divergent and convergent thinking are directly correlated with research impact. Here, we think of divergent and convergent thinking as coupled (interactive) processes, and that the quality and quantity of the individual process as well as the interactivity between the processes determine the quality and quantity of the end products: ideas (new knowledge). The higher the quality, quantity and interactivity of the underlying processes, the higher the quality and quantity of the new knowledge and the higher the quality and quantity of the research impact.

Third: divergent and convergent thinking, like other aspects of research (Narayanamurti et al. 2009), occur throughout the research ecosystem (Hueske et al. 2015). In using the phrase ‘research ecosystem’, we deliberately make the metaphor to ‘biological ecosystem’ and hence to the importance of the multiple levels of such an ecosystem: individuals; groups of individuals; and the environment which sets the reward/cultural boundary/interaction conditions for the individuals and groups of individuals. These three levels also map to the micro-, meso- and macro-scales identified by the Science of Team Science (SciTS) community (Börner et al. 2010). Moreover, divergent and convergent thinking can be inhibited at all of these levels by various ‘inhibitory mechanisms’. The mechanisms can be: cognitive and operating at the level of the individual researcher; social and operating at the level of the research team; or cultural/organizational and operating at the level of the research institution.

Complementary to these three overarching hypotheses are three overarching science opportunities. First, can we measure divergent and convergent thinking? Are there signatures in the thought and communication patterns—of individuals, of teams and of individuals and teams across an institution—that can be associated with the two kinds of thinking? Second, how can divergent and convergent thinking be correlated with research impact? Third, what are the most important mechanisms by which divergent and convergent thinking are inhibited, and are there dis-inhibitory interventions?

The remainder of the chapter is an enumeration of examples of (a) inhibitory mechanisms, which in our experience as physical science and engineering researchers commonly inhibit divergent and convergent thinking at the three levels of the research ecosystem (individuals, teams, institution), along with a brief description of the foundational social science principles underlying those mechanisms and (b) dis-inhibitory interventions (frequently based in new technologies) that might help neutralize those inhibitory mechanisms. Our first goal with these examples is to clarify, and to make plausible, the centrality of divergent and convergent thinking processes to research. Our second goal is to illustrate how common in current practice inhibitory mechanisms are in decreasing the quality and quantity of divergent and convergent thinking. Our third goal is to catalyse new work on these inhibitory mechanisms—to understand how they operate, how they might be measured scientifically and how they might be neutralized through various dis-inhibitory interventions, especially new approaches that are at least partially based in emerging technologies.

14.3 Individual Researchers: Human Cognitive Constraints and Biases

At the individual researcher level, let us consider first divergent thinking, second convergent thinking and third balancing the two kinds of thinking.

14.3.1 Divergent Thinking: Overcoming Idea Fixation Through Engineered Exposure to New Ideas

Divergent thinking, in essence, is the creation of new ideas, mostly, perhaps always (Arthur 2009), through the recombination of pre-existing ideas. However, humans have cognitive constraints and biases (Kahneman 2011), which can make divergent thinking difficult. Prominent among these is idea fixation, an inability to break free from ideas that preoccupy the mind and hold attention (Linsey et al. 2010). In our experience, such idea fixation is a common inhibitory mechanism to effective divergent thinking.

Nonetheless, productive researchers must and do de-fixate themselves at key stages of their research process. In many cases, the de-fixation takes place through serendipitous exposure to new ideas. A famous example is Charles Darwin's exposure to the ideas of Thomas Malthus (1826), of which he writes in his autobiography 'In October 1838, fifteen months after I had begun my systematic inquiry, I happened to read for amusement Malthus on Population, and being prepared to appreciate the struggle for existence which everywhere goes on, from long-continued observation of the habits of animals and plants, it at once struck me that under these circumstances favourable variations would tend to be preserved, and unfavourable ones to be destroyed. The result would be the formation of a new species'. (Darwin 1887). Indeed, highly creative institutions value serendipitous exposure to new ideas so highly that they sometimes try to probabilistically enhance informal interactions through engineered physical spaces, e.g.: MIT's Building 20 (Lehrer 2012), Bell Labs' 'Infinite Corridor' (Tsao et al. 2013), the Janelia Farm Research Campus (2003), Pixar's Emeryville campus (Catmull and Wallace 2014) and Las Vegas' Downtown Project (Singer 2014). Break activities such as food in shared spaces can also enhance the cross section for such informal interactions (Emmanuel and Silva 2014).

The engineering of physical spaces to enhance serendipitous exposure to new ideas is thus a possible dis-inhibitory intervention that can help neutralize idea fixation. Here, we propose that engineered exposure to new ideas might be another and more direct possibility. The trick is that the new ideas should be far enough away in analogical space (Chan et al. 2015; Fu et al. 2013) to catalyse shifts in perspective—either because they come from different problem areas of the same discipline, from different disciplines or from different 'translational' (science, technology, applications) communities. The ideas should not be so far away in analogical space, however, that conceptual and language gaps are too difficult to bridge.

Such an 'engineering exposure to optimal-analogic-distance ideas' dis-inhibitory intervention is certainly practiced in a qualitative way by experienced research managers when they see their staff 'stuck'. However, advances in modern data analytics, combined with the sheer quantity of digitized knowledge, open up new opportunities for making this practice more quantitative. One opportunity might be scientometric clustering analyses of publications based on bibliographic connectivity. Another opportunity might be lexical clustering analyses based on syntactic/semantic regularities (Mikolov et al. 2013), word-order-based discovery of underlying ('latent') constituent topic areas (Blei et al. 2003) and mutual compressibility (Cilibrasi and Vitányi 2005). These analyses could lead to algorithms that go beyond those that power today's search (Salton 1975) and recommendation (Bennett et al. 2007) engines, by feeding researchers and engineers ideas not just within their comfort zone, but optimally distant from their comfort zone (Fu et al. 2013).

14.3.2 Convergent Thinking: Overcoming Sloppy Thinking Through Disciplined Use of Research Narratives

Convergent thinking, in essence, is the testing and selection from newly generated ideas those worth pursuing, ideally through logic and analysis. As Linus Pauling once said (Mulgan 2006): 'the way to get good ideas is to get lots of ideas and throw the bad ones away'. Of course, easier said than done, because human cognition is subject to sloppy thinking and errors of logic and analysis. Logic and analysis require hard work, and with finite time and resources, humans have evolved to use heuristics to make decisions which 'satisfice' (Simon 1956). In research, however, knowledge builds on knowledge, incorrect knowledge can misguide and negate a pyramid of subsequent research and the balance between heuristics and logic/analysis must be shifted away from heuristics and towards logic/analysis.

There is thus opportunity for understanding the cognitive science basis for the many heuristics that are essentially inhibitory mechanisms for effective convergent thinking, and for developing dis-inhibitory interventions that enable or force those heuristics to be side-stepped. Of particular interest is an intervention that might be called the 'research narrative' intervention. Research narratives—storylines which knit together background, hypothesis, methodology, analysis, findings and implications—are essentially tools for logical thinking. Mathematics is perhaps the highest form of such logical thinking, and in our experience as physical scientists, it can be applied at various research stages to overcome sloppy thinking.

Research narratives are obviously important at the end of a research project, when a paper is being written for the scientific community and posterity. It is at this stage that many loose ends are discovered that cause a revisiting of the work itself, or at least of the interpretation of the work. Antoine Lavoisier, for example, famously did not 'discover' the role of oxygen in combustion until he began to piece together the research narrative associated with his experiments (Cole 1992). And Paul Dirac, for example, did not 'discover' antiparticles until he derived and then followed to its logical conclusion of his equation that combined quantum theory and special relativity to describe the behaviour of an electron moving at relativistic speeds (Dirac 1928).

But research narratives are just as important at the beginning of a project. Emerging cognitive science suggests that narrative and stories are the evolutionary optimal tools for communicating not only with others but even with ourselves (Gottschall 2012). A coarse storyboard of the title, abstract, figures and key references of the anticipated outcome of a project forces clarification of many of its aspects—including those that have been hypothesized (Simonton 2013) to be critical sub-components of creative ideas, such as originality, perceived utility and 'surprisingness'. A project whose narrative does not hang together is in danger of Wolfgang Pauli's famously blistering criticism: 'What you said was so confused that one could not tell whether it was nonsense or not' (Peierls 1960).

Not everyone is 'good' at crafting or even self-assessing their own research narratives, though. There is thus potential for the research narrative intervention to be augmented by modern data analytics. For example, computer algorithms could dispassionately evaluate research narratives just as they are now dispassionately evaluating essays in academic writing courses (Foltz et al. 1999). Or, perhaps more likely, a *combination* of machines and humans might someday efficiently and accurately evaluate research narratives via machine curation of Yelp-like peer reviews.

Note that this research narrative dis-inhibitory intervention enables one to see the close interplay between divergent and convergent thinking. As one creates a research narrative, one often finds that the conclusion one anticipates is not supported by the narrative; instead, as in the Lavoisier example above, another conclusion is. When the research narrative makes heavy use of mathematics this happens often, as in the Dirac example above: mathematics or simulations, even when guided by a conclusion that intuition has presaged, often leads instead to a different and surprising conclusion. The research narrative has eliminated one conclusion, thus enhancing the effectiveness of a convergent thinking, but it has also unearthed a possible new conclusion, thus enhancing the effectiveness of divergent thinking. Many researchers are effective at doing one precisely because they are so effective at doing the other: those who can, in their heads, analyse and eliminate bad ideas (i.e. who are good at convergent thinking), can free themselves from keeping those ideas in their mind, and can spend more time thinking up new ideas (divergent thinking).

14.3.3 Balancing Divergent and Convergent Thinking

Divergent and convergent thinking are by themselves difficult, but perhaps even more difficult is our ability to know when to switch between the two.

On a large scale, the history of science and innovation is replete with scientists and engineers who were on the wrong track, and would have been more productive switching from convergent to divergent thinking (Isaacson 2007), at least in terms of the problem spaces they were considering. Albert Einstein famously spent the last 30 years of his life on a fruitless quest for a way to combine gravity and electromagnetism into a single elegant 'unified field' theory. But the history of science and innovation also has its share of scientists and engineers who abandoned or postponed attacking certain problem spaces, which later they thought they should have attacked. As Edwin Jaynes, a physicist who made many fundamental contributions to statistical mechanics and Bayesian/information theory poignantly expressed it: 'Looking back over the past forty years, I can see that the greatest mistake I made was to listen to the advice of people who were opposed to my efforts. Just at the peak of my powers I lost several irreplaceable years because I allowed myself to become discouraged by the constant stream of criticism from the Establishment, that descended upon everything I did [...]. The result was that my contributions to probability theory were delayed by about a decade, and my potential contributions to electrodynamics—whatever they might have been—are probably lost forever' (Jaynes 1993). Thus there is a critical dilemma: when researchers are being confronted with something unexpected, they must choose whether to stay the course (convergent thinking) or to treat the unexpected as an opportunity to reconsider possibilities (divergent thinking).

To some extent, people gravitate towards thinking styles with which they are most comfortable, and researchers are no different. Those who are more comfortable thinking divergently will tend to reconsider too soon; those more comfortable thinking convergently will tend to stay the course too long; and perhaps a rare few will be comfortable doing neither [as in F. Scott Fitzgerald's famous quote 'The test of a first-rate intelligence is the ability to hold two opposed ideas in the mind at the same time, and still retain the ability to function' (Fitzgerald 2009)]. In fact, because of our modern education system's emphasis on deducing single answers using logical thinking, modern researchers might be biased towards convergent thinking. To avoid this bias, some institutions that value creativity now deliberately hire on the basis not of grade point average (GPA) and scholastic aptitude test (SAT) scores, but of more balanced thinking styles (D'Onfro 2014).

There is thus opportunity to understand and engineer strategies to compensate for intrinsic biases towards either divergent or convergent thinking. For example, at a qualitative level, the research narratives discussed earlier might not just be powerful tools for logical, convergent thinking, but might also be powerful tools for understanding when to cycle between divergent and convergent thinking. If the train of thought that follows from one or more ideas does not hold up to the cold logic (or mathematics) of the research narrative, then very likely new ideas and divergent thinking are needed.

Or, for example, at a quantitative level, some of the lexical analytical techniques mentioned earlier, applied in real time to evolving research narratives and other generated knowledge trails, might be able to discover not only whether divergent or convergent thinking is happening, but whether divergent or convergent thinking is *appropriate* for the stage of the problem at hand. In the language of chemical engineering, one would like to understand which kind of thinking is rate limiting, and therefore which to focus attention on, at a given instant in time.

14.4 Research Teams: Social Constraints and Biases

At the research team level, let us similarly consider first divergent thinking, second convergent thinking and third with balancing the two.

14.4.1 Divergent Thinking: Overcoming Over-Reliance on Strong Links by Exploiting Weak Links

First, we state the obvious: research teams have the potential to think much more divergently than individuals can. Groups can draw upon the diverse ideas of individuals to create new ideas. And, because much of the knowledge of individuals is tacit (Polanyi 1967) and not accessible in formal codified form, closely interacting groups which can share this tacit knowledge informally can be yet more productive.

However, research teams also bring inefficiencies to divergent thinking. When individuals on a team become too strongly socially linked to each other and have become familiar with each other's knowledge domains and ways of thinking, they no longer serve as sources of new ideas to each other. Moreover, homophily is common in social networks: we seek those who think as we do and avoid those who do not think as we do (McPherson et al. 2001). For divergent thinking, exposure to the less familiar is important, and thus weak links (Granovetter 1973) in one's social network can be more powerful than strong links. Creativity has been found to be greatest when the relationship distance between collaborators is intermediate—neither too close nor too far. For example, when using statistical methods to examine the success of Broadway musical artists, Uzzi and Spiro (2005) found a relationship that indicated lack of creativity when artists worked in a social network that was overly familiar or overly unfamiliar to them.

Thus, an over-reliance on strong links in one's social network, or an over-reliance on the members of one's research project team, can be thought of as an inhibitory mechanism to effective divergent thinking at the research team level. The obvious dis-inhibitory intervention is to exploit weak links in the team's social network, and to expose members of the research project team to weakly linked new people who bring new ideas. This exposure to new people can of course be done serendipitously, as discussed above at the individual researcher level. More interestingly, it can also be self-engineered. A famous example is the Wright Brothers, who formed their own nuclear team, but were also in constant contact with like-minded experts from all over the world whose ideas they incorporated into their own airplane designs (Sawyer 2011).

But now the exploitation of weak links can also be externally engineered, through the use of data analytics to identify not just ideas that are an optimal analogic distance away from the current team's ideas, but people who are an optimal analogic distance away from people in the current team. Because social trust helps foster communication, one might even imagine biasing the matchmaking towards people who are not just an optimal analogic distance away, but are also socially linked in some fashion to the existing team. This would be similar to the matchmaking of people for romantic purposes, but here for the purpose of optimal idea cross-fertilization. The world is a big place, which means that it must be exploited but also that it is difficult to exploit, so engineered links are key.

14.4.2 Convergent Thinking: Balancing Impermeable Teams with Permeable Collaborations

Just as for divergent thinking, research teams can be much more effective at convergent thinking than individuals can be. Convergent thinking requires logical deductive thinking, the deeper and more first-principled the more accurate and often the more surprising (Lucibella and Blewett 2013). In studies of lab groups, other researchers in the lab group regularly found and fixed reasoning errors made by individual researchers (Dunbar 1997). Further, multiple minds can span the expertises required to rigorously test ideas. Consider for example the capabilities contained in the Joint Center for Energy Storage Research, a $120 M (over 5 years) effort to achieve revolutionary advances in battery performance. Because the work spans chemistry, materials science, physics, computational theory and nanoscience, it would be impossible for an individual researcher to span these areas.

There is overhead, however, associated with research teams that contain multiple capabilities and expertises: the teams must contain the *right* capabilities and expertises. If they do not, then the effectiveness of idea testing and selection (convergent thinking) will be severely compromised. It is not uncommon in our experience for a team to be addressing a problem or testing an idea using very cleverly the capabilities and expertises that exist within the team, but not using a capability or expertise that could be extremely helpful but does not exist within that team. The team may of course be unaware of the helpful outside capability or expertise (you do not know what you do not know), but just as often it is aware but does not have the flexibility to reconfigure itself so as to add the outside capability or expertise. The team might have been formed at a time when the problem required a certain set of capabilities and expertise, but the problem has evolved and now the necessary capabilities and expertises are different. Because of funding constraints and/or social glue/loyalty, it proves difficult to add people to and subtract people from the team as needed. This *impermeability* of teams to composition reconfiguration is in our experience a common inhibitory mechanism to effective convergent thinking: teams that are highly impermeable pay a non-agility cost and are less able to accommodate in real time quickly evolving goals and approaches.

The logical dis-inhibitory intervention is to rebalance away from impermeable teams and towards permeable collaborations. Self-organized collaborations, in which researchers choose on their own with whom to collaborate for the purpose of solving a particular problem of interest at a particular instant in time, can evolve quickly as new problems arise. Paul Erdös, the most prolific mathematician in history, was famous for his peripatetic always-on-the-lookout-for-the-next-problem-and-the-next-collaborator style: 'Erdös structured his life to maximize the amount of time he had for mathematics […]. In a never-ending search for good mathematical problems and fresh mathematical talent, Erdös crisscrossed four continents at a frenzied pace, moving from one university or research center to the next. His modus operandi was to show up on the doorstep of a fellow mathematician, declare, "My brain is open", work with his host for a day or two, until he was bored or his host was run down, and then move on to another home'. (Hoffman 1998).

Understanding how to balance impermeable teams and permeable collaborations is non-trivial, however. The question is analogous to Ronald Coase's famous economics puzzle—when do we need firms as opposed to free agents interacting in a marketplace? Coase's answer had to do with transaction costs: firms (and, for us, research teams) reduce certain kinds of transaction costs amongst the employees of the firm that are difficult to reduce when people just interact with each other as completely free agents. Perhaps the most important transaction cost is the building of trust—trust that your firm-mates (and for us, research teammates) will get done what they promise to get done, trust that you have a job and retirement security without having to always be fending for yourself, trust in a basic social safety net.

If we take this point of view, then we need teams when impermeability and hierarchy are a net plus—when you need a division of labour to execute efficiently and when the problem/solution spaces are relatively mature hence enabling teams to move forward productively without a need for continual reconfiguration. And you do not want fixed teams when permeability and self-organization are a net plus—when you need agility in idea generation/test and when the problem/solution spaces are relatively immature for the problems at hand and still evolving rapidly.

A major challenge and opportunity are thus to learn how to match the right degree of impermeability/permeability with the maturity/immaturity of the research problem space being attacked. This will require learning how to measure the degree of impermeability or permeability that characterize research teams and collaborations, learning how to measure the degree of maturity or immaturity of the research problem space and then learning how the match or mismatch between the two correlates with research success. Emerging sociometric tools based on physical and electronic communications (Waber 2013) will likely play a role in effective measures of permeability, and data harvested from project management software could also be used to look across teams to understand relative maturity of the research.

14.4.3 Distributing Divergent and Convergent Thinking Between Individuals and Teams

Most importantly, research teams have more options for accomplishing divergent and convergent thinking than do individuals. Teams are *composed* of individuals. Hence, if some aspect of thinking is best done by a team or by individuals, teams can in principle assign it to the appropriate level. For example, if individuals are relatively stronger at convergent thinking while teams are relatively stronger at divergent thinking, it could be optimal for divergent thinking to be performed more at the team level, but for convergent thinking to be performed more at the individual level (Shore et al. 2014). To take advantage of this strategy, however, it will be necessary to first understand more deeply the relative strengths and weakness of individuals and teams at convergent and divergent thinking for what types of problems, in what situations and environments and using what interaction tools.

Teams also have more options in how their individual members are rewarded. Individual researchers not in a team would individually bear the consequences of risky too-divergent thinking, but in a team could actually be rewarded for taking on such risk. An example from nature is scout ants, who individually have a high rate of demise but their ‘divergent thinking’ is important to the colony because it helps them work together in innovative ways to provide survival for current and emerging colonies (Wilson 2012).

Countering the above, research teams have fewer options for oscillating back and forth between divergent and convergent thinking during the life cycle of a research project. They inherently have more inertia, and thus the decision of what kind of thinking to emphasize and at what level, individual or team, is more serious.

For all the above reasons, team leadership is crucial. Throughout the life cycle of a project, a team will move through various quadrants of individual/team divergent/convergent thinking, with opportunity for the team and its leader to optimally allocate resources across those quadrants. As mentioned in the Introduction, successful leaders of research teams have developed a deep understanding of how to do this won from hard-earned experience. However this understanding is intuitive, without the tools necessary for quantitative analysis. Thus, we call here for more measurements and greater use of modern data analytics to analyse those measurements. For example, can we quantify: where in its life cycle a research project is; the degree to which divergent or convergent thinking is needed and how well the team's current composition and cognitive constructs (Dong 2005) match the desired degree of divergent or convergent thinking? Just as the 'quantified self' movement (Swan 2013) seeks to use physical technology to monitor the manifold pulses of a person's daily life to optimize health and productivity, a 'quantified research team' movement might seek to use data analytics technology to monitor the manifold pulses of a research team's daily life (Waber 2013), to better match the team's composition and organization to the research challenge at hand, and ultimately to optimize the team's health and productivity.

Understanding how to optimize the balance between individual and team, and between divergent and convergent thinking, might also borrow from advances in emerging models for information foraging (Pirolli and Card 2005). For example, if useful information is 'patchy', a forager might first seek to look broadly for useful patches, and then focus in on a few of the most useful patches. Or, for example, the risk associated with not finding a patch in a particular time horizon, or the amount of resources allocated for the foraging, might determine which stage of the foraging is best done by individuals or by a team.

14.5 Research Institutions: Cultural Constraints and Biases

At the research institution level, there are outsized opportunities for optimization, because it is this level that defines the culture and reward system within which individual researchers and research teams are drawn from and within which they engage in divergent and convergent thinking.

14.5.1 Divergent Thinking: Balancing a Culture of Performance with a Culture of Learning

Divergent thinking, whether by individual researchers or research teams, takes place within an institutional culture. Institutional culture, in turn, evolves so that the institution self-preserves—those institutions whose cultures do not evolve in this manner go extinct, leaving behind those whose cultures did evolve in this manner. Since the quickest way for an institution to not self-preserve is to not meet its existing commitments, performance to those commitments almost always becomes a central part of its culture. A culture of conservativism—making commitments which are conservative, then meeting those commitments through approaches which are conservative—is natural. This can be the case even for research institutions that aim to be at the forefront of knowledge production, hence ostensibly tackling tough and perhaps intractable problems that require new ideas and divergent thinking.

Using the language popularized by Dweck and Leggett (1988), the culture rewards a 'performance' rather than a 'learning' mindset. By a performance mindset we mean a mindset that focuses on the immediate outcome or achievement as an external judgment on abilities which are perceived to be 'fixed'. By a learning mindset we mean a mindset that focuses not so much on success or failure but on the opportunity to learn and to enhance abilities which are perceived to be capable of 'growth'. As Mevin Kelley, former Director of Research at Bell Labs who hired Shockley, Bardeen, Baker and many other great scientists and engineers, would say to new recruits: 'Ask not what you know, but what you don't know' (Narayanamurti 2016).

A culture of conservatism and performance is thus a strong mechanism for inhibiting divergent thinking, for avoiding the risks associated with such thinking, and most importantly for avoiding those risks throughout the sequence of problem and solution spaces that are typically explored in research. First, there is the initial problem space: what is the most interesting, challenging yet potentially solvable problem to propose to tackle in the first place? Conservatism dictates proposing too-safe problems whose solutions are already in hand. Second, there is the solution space: what is the deepest and most elegant solution with the widest fan-out even beyond the immediate problem at hand? Conservativism dictates proposing too-safe solutions that solve the immediate problem but little else. Third, there is the revisit of the problem space: as the work proceeds, are there even more important alternative problems that intermediate results promise to solve, possibly with some deviation from the original work plan? Conservativism dictates not deviating from the original work plan and blinding oneself to alternative problems. In other words, the culture of performance inhibitory mechanisms affects divergent thinking insidiously in both the solution *and* problem spaces.

A well-known example of how valuable it can be to expand the problem space to accommodate solutions that do not match the original problem space is 3M's discovery of its Post-It Notes. Its origin was in the desire to create super-strong adhesives for use in the aerospace industry. A super-weak but pressure-sensitive adhesive was created enroute and was initially dismissed. But by an expansion of the problem space beyond the aerospace industry into consumer office supply market, the new adhesive, albeit with some deviation from the original work plan, became the foundation for one of the best-selling office supply products ever. Even in a culture as biased towards learning as 3M's, however, the twists and turns of this expansion of problem space were considerable and Post-It Notes barely made it to market. In a culture biased towards performance, it would have been that much less likely that Post-It Notes would have come to market.

The obvious dis-inhibitory intervention to a culture that overemphasizes performance is to rebalance towards a culture of learning. As mentioned above, this rebalancing is in a direction opposite to that which is enabling the institution as a whole to preserve itself in the short term. However, it is in a direction consistent with that which would enable the institution as a whole to preserve itself in the long run, and thus a direction which enlightened management might pursue. The institution could do this by rewarding reasonable risk-taking and dis-rewarding unreasonable risk aversion, so as to reward the choosing of optimally challenging problems and solutions. Tools can be created for assessing and managing risk portfolios in research just as they exist in large numbers in financial planning. An institution that focuses on standard academic metrics such as GPA and SAT scores when hiring is selecting for strength in convergent rather than divergent thinking (Chamorro-Premuzic 2006). New meausures should be added that also assess excellence in divergent thinking (D'Onfro 2014). The history of science and innovation is replete with extremely productive researchers who did not do well in school, and is also replete with less productive researchers who did extremely well in school: in research divergent thinking is just as necessary as convergent thinking.

14.5.2 Convergent Thinking: Balancing a Culture of Consensus with a Culture of Truth

Convergent thinking also takes place within an institutional culture. As discussed just above, though institutional culture tends to value convergent over divergent thinking, there are nonetheless many ways in which institutional culture can also inhibit effective convergent thinking. One that we see often in our personal experience is what might be called a culture of consensus. This is a culture that values social and intellectual harmony and punishes social and intellectual disharmony. Such biases towards harmony are easy to understand and probably evolved in humanity's prehistory for good reason: there are many situations for which quick consensus, conflict avoidance and social cohesion are more important than accuracy. Those situations likely do not include among them research, for which there is a higher importance placed on accuracy. Consensus that is socially driven and not the beneficiary of deep intellectual debate and intermediate stages of disharmony does not achieve truth but instead groupthink, in which groups converge prematurely and inaccurately on false or less-good ideas (De Dreu et al. 1999).

An institutional culture of consensus is thus a strong inhibitory mechanism to effective convergent thinking. The obvious dis-inhibitory intervention is to rebalance towards what might be called a culture of truth. Such a culture is non-trivial to achieve, of course, because truth and the path to truth is uncomfortable. But a great research culture is probably *not* one in which individuals are 'comfortable'. Truth requires individuals and teams to go beyond their intellectual comfort zones into Kuhn's 'essential' (Kuhn 2012) or Senge's 'creative' (Senge and Suzuki 1994) tension; divergent and convergent thinking do not so much cause comfort as they cause stress. Indeed, just as one would expect elite athletes and athletic teams to experience significant levels of stress in the game, indiviuals and research teams are also strained by the demands of innovative research solutions.

Not all stress is created equal, however. Cognitive and intellectual stress is essential, and one might argue the more the better. Existential (the threat of losing one's job or funding) and social (the threat of losing one's social standing) stress is not directly essential, and one might argue the less the better. The opportunity here is thus twofold. A first opportunity is to understand how to measure these two types of stress. For example, could text analytics (Pennebaker 2014) extract the magnitudes of the two types of stress either from single emails or from massive numbers of internal emails within an institution, thereby 'taking the pulse' of the institution? A second opportunity is to engineer changes in the research environment to maximize the first, and minimize the second, type of stress. Here, though, one might imagine two possibilities.

On the one hand, if existential/social stress were independent of cognitive/intellectual stress, one might seek to simply 'zero out' the first, e.g. by creating 'secure bases' (Bowlby 2005) for research funding and social rewards oriented towards teams and not individuals. One might make more disciplined use of strategies [including the Delphi method or its variants (Okoli and Pawlowski 2004) and certain kinds of analogy use (Paletz et al. 2013)] for enhancing the task or technical conflict necessary for unveiling logical inconsistencies, while minimizing the social conflict whose avoidance would otherwise foster groupthink. If, as has been speculated (Buchen 2011), those with some degree of Asperger's, who are less aware of and less sensitive to social cues, are also least susceptible to groupthink and thus often the deepest and most first-principles thinkers, one might imagine a more disciplined harnessing of these thinkers (Silberman 2015).

On the other hand, if existential/social stress is not independent, but is an inseparable motivator, of cognitive/intellectual stress, then one might devise strategies which harness existential/social stress while controlling its degree. For example, just as some aspects of war can be gamified (Macedonia 2002), some aspects of research might also be gamified (Deterding et al. 2011; McGonigal 2011)—thus harnessing, but not letting get out of hand, existential/social stress.

14.6 A Vision for the Future

Research, the manufacture of knowledge and products, is complex and practiced largely as an art. As science (understanding) and technology (tools) continue to be developed and applied to the manufacturing of those other goods and services, it is natural, perhaps inevitable, that they will also be applied to research. From our perspective as physical scientists, we call here for this application to begin seriously in the physical sciences just as it has already begun in the life sciences.

Drawing on a recent workshop of EPS researchers/managers and social scientists of science, we have posited in this chapter a framework in which the microscopic processes underlying research are iterative and closely interacting cycles of divergent (generation of ideas) and convergent (testing and selecting of ideas) thinking processes. We anticipate that an improved understanding of these microscopic processes, and of how various 'inhibitory mechanisms' can prevent them from being executed effectively, can ultimately help us design and engineer appropriate 'dis-inhibitory interventions'. Among the commonplace inhibitory mechanisms we identified, along with corresponding potential dis-inhibitory interventions, were: overcoming idea fixation through engineered exposure to new ideas; overcoming sloppy thinking through disciplined use of research narratives; overcoming over-reliance on strong links by exploiting weak links; balancing impermeable teams with permeable collaborations; balancing a culture of performance with a culture of learning and balancing a culture of consensus with a culture of truth.

Finally, we have focused in this chapter only on the direction from 'science to art', in which the emerging science of research is harnessed to improve the art of research. Ultimately, even greater opportunity will be unleashed when the other direction from 'art to science' is also exercised simultaneously and synergistically—when improvements in how research is actually done are used to test our understanding of how research is done. Within the physical sciences, such a close and bidirectional interaction between science (understanding) and art (technology and tools) has resulted in well-documented spirals of mutual benefit (Brooks 1994; Casimir 2010; Narayanamurti et al. 2013; Tsao et al. 2008). One can anticipate close bidirectional interaction between the science and art of research to result in similar spirals of mutual benefit. If we narrow the type of research to be examined and improved to the physical sciences and engineering, the analogy would be to Asimov's Foundations 2 and 1 (Asimov 2012), except, rather than warring, engaging in a complementary partnership.

Acknowledgements GEA, ARS and JYT acknowledge support from Sandia National Laboratories. Sandia National Laboratories is a multi-mission laboratory managed and operated by National Technology & Engineering Solutions of Sandia, LLC, a wholly owned subsidiary of Honeywell International, Inc., for the U.S. Department of Energy's National Nuclear Security Administration under contract DE-NA0003525.

References

Arthur, W. B. (2009). *The nature of technology: What it is and how it evolves*. New York, N.Y.: Simon and Schuster.

Asimov, I. (2012). *Foundation's edge*. New York, N.Y.: Random House LLC.

Bennett, J., Lanning, S., & Netflix, N. (2007). *The Netflix prize*. Paper presented at the In KDD Cup and Workshop in conjunction with KDD.

Beveridge, W. I. B. (1957). *The art of scientific investigation*. New York, N.Y.: WW Norton & Company.

Blei, D. M., Ng, A. Y., & Jordan, M. I. (2003). Latent dirichlet allocation. *The Journal of Machine Learning Research, 3,* 993–1022.

Börner, K., Contractor, N., Falk-Krzesinski, H. J., Fiore, S. M., Hall, K. L., Keyton, J. … Uzzi, B. (2010). A multi-level systems perspective for the science of team science. *Science Translational Medicine, 2*(49), 49cm24–49cm24.

Bowlby, J. (2005). *A secure base: Clinical applications of attachment theory* (Vol. 393). UK: Taylor & Francis.

Brooks, H. (1994). The relationship between science and technology. *Research Policy, 23*(5), 477–486.

Buchen, L. (2011). When geeks meet. *Nature, 479*(7371), 25–27.

Campbell, D. T. (1960). Blind variation and selective retentions in creative thought as in other knowledge processes. *Psychological Review, 67*(6), 380.

Casimir, H. B. G. (2010). *Haphazard reality: Half a century of science*. Netherlands: Amsterdam University Press.

Catmull, E., & Wallace, A. (2014). *Creativity, Inc.: Overcoming the unseen forces that stand in the way of true inspiration*. New York, N.Y.: Random House LLC.

Chamorro-Premuzic, T. (2006). Creativity versus conscientiousness: Which is a better predictor of student performance? *Applied Cognitive Psychology, 20*(4), 521–531.

Chan, J., Dow, S. P., & Schunn, C. D. (2015). Do the best design ideas (really) come from conceptually distant sources of inspiration? *Design Studies, 36,* 31–58. https://doi.org/10.1016/j.destud.2014.08.001.

Cilibrasi, R., & Vitányi, P. M. (2005). Clustering by compression. *Information Theory, IEEE Transactions on, 51*(4), 1523–1545.

Cole, S. (1992). *Making science: Between nature and society*. USA: Harvard University Press.

Cropley, A. (2006). In praise of convergent thinking. *Creativity Research Journal, 18*(3), 391–404.

D'Onfro, J. (2014, July 12, 2014). *Here's why Google stopped asking bizarre, crazy-hard interview questions*. Retrieved from http://www.businessinsider.com/google-hiring-practices-interviews-2014–7.

Damanpour, F. (1991). Organizational innovation: A meta-analysis of effects of determinants and moderators. *Academy of Management Journal, 34*(3), 555–590.

Damanpour, F., & Gopalakrishnan, S. (2001). The dynamics of the adoption of product and process innovations in organizations. *Journal of Management Studies, 38*(1), 45–65.

Darwin, C. (1887). *The Autobiography of Charles Darwin*. Barnes & Noble Publishing.

De Dreu, C. K., De Vries, N. K., Gordijn, E. H., & Schuurman, M. S. (1999). Convergent and divergent processing of majority and minority arguments: Effects on focal and related attitudes. *European Journal of Social Psychology, 29*(23), 329–348.

Deterding, S., Dixon, D., Khaled, R., & Nacke, L. (2011). *From game design elements to gamefulness: Defining gamification.* Paper presented at the Proceedings of the 15th International Academic MindTrek Conference. Envisioning Future Media Environments.

Dirac, P. A. M. (1928). The quantum theory of the electron. *Proceedings of the Royal Society A: Mathematical, Physical and Engineering Sciences, 117*(778), 610-–624.

Dong, A. (2005). The latent semantic approach to studying design team communication. *Design Studies, 26*(5), 445–461.

Drucker, P. F. (2011). *Technology, management, and society.* USA: Harvard Business Press.

Dunbar, K. (1997). How scientists think: On-line creativity and conceptual change in science. In T. B. Ward & S. M. Smith (Eds.), *Creative thought: An investigation of conceptual structures and processes* (pp. 461–493). Washington, D.C., USA: American Psychological Association.

Dweck, C. S., & Leggett, E. L. (1988). A social-cognitive approach to motivation and personality. *Psychological Review, 95*(2), 256.

Emmanuel, G., & Silva, A. (2014). Connecting the physical and psychosocial space to Sandia's mission. *Sandia Report*, SAND2014-16421.

Fealing, K. (2011). *The science of science policy: A handbook.* USA: Stanford University Press.

Feist, G. J. (2008). *The psychology of science and the origins of the scientific mind.* USA: Yale University Press.

Fitzgerald, F. S. (2009). *The crack-up*. USA: New Directions Publishing.

Foltz, P. W., Laham, D., & Landauer, T. K. (1999). *Automated essay scoring: Applications to educational technology.* Paper presented at the World Conference on Educational Multimedia, Hypermedia and Telecommunications.

Fu, K., Chan, J., Cagan, J., Kotovsky, K., Schunn, C., & Wood, K. (2013). The meaning of "near" and "far": The impact of structuring design databases and the effect of distance of analogy on design output. *Journal of Mechanical Design, 135*.

Fu, K., Chan, J., Schunn, C., Cagan, J., & Kotovsky, K. (2013b). Expert representation of design repository space: A comparison to and validation of algorithmic output. *Design Studies, 34*(6), 729–762. https://doi.org/10.1016/J.Destud.2013.06.002.

Gottschall, J. (2012). *The storytelling animal: How stories make us human*. USA: Houghton Mifflin Harcourt.

Granovetter, M. S. (1973). The strength of weak ties. *American Journal of Sociology*, 1360–1380.

Grueber, M., & Studt, T. (2014). 2014 global R&D funding forecast. *R&D Magazine, 16,* 1–35.

Haak, L. L. (2014, Mary 24, 2014). *A vision to transform the research ecosystem*. Retrieved from http://www.editage.com/insights/a-vision-to-transform-the-research-ecosystem.

Hoffman, P. (1998). *The man who loved only numbers*. New York: Hyperioncop.

Hueske, A. K., Endrikat, J., & Guenther, E. (2015). External environment, the innovating organization, and its individuals: A multilevel model for identifying innovation barriers accounting for social uncertainties. *Journal of Engineering and Technology Management, 35,* 45–70.

Isaacson, W. (2007). *Einstein: His life and universe*. New York, N.Y.:Simon and Schuster.

Janelia Farm Research Campus: Report on Program Development. (2003). Retrieved from http://www.janelia.org/sites/default/files/JFRC.pdf.

Jaynes, E. T. (1993). A backward look to the future. *Physics and probability*, 261–275.

Jones, C. I. (1995). R & D-based models of economic growth. *Journal of political Economy*, 759–784.

Kahneman, D. (2011). *Thinking, fast and slow*. UK: Macmillan.

Kuhn, T. S. (2012). *The structure of scientific revolutions*. USA: University of Chicago press.

Lehrer, J. (2012, January 30, 2012). Groupthink: The brainstorming myth. *The New Yorker*.

Linsey, J., Tseng, I., Fu, K., Cagan, J., Wood, K., & Schunn, C. (2010). A study of design fixation, its mitigation and perception in engineering design faculty. *Journal of Mechanical Design, 132* (4), 041003.

Lucibella, M., & Blewett, H. (2013, October, 2013). Profiles in versatility: Part 1 of two-part interview: Entrepreneur Elon Musk talks about his background in physics. *APS News, 22*.

Macedonia, M. (2002). Games soldiers play. *Spectrum, IEEE, 39*(3), 32–37.

Malthus, T. R. (1826). *An Essay on the Principle of Population; Or, A View of Its Past and Present Effects on Human Happiness: With an Enquiry Into Our Prospects Respecting the Future Removal of Mitigation of the Evils which it Occasions*. John Murray.

March, J. G. (1991). Exploration and exploitation in organizational learning. *Organization Science, 2*(1), 71–87.

McGonigal, J. (2011). *Reality is broken: Why games make us better and how they can change the world*. Penguin.

McPherson, M., Smith-Lovin, L., & Cook, J. M. (2001). Birds of a feather: Homophily in social networks. *Annual review of sociology*, 415–444.

Medawar, P. B. (1963). Is the scientific paper a fraud. *The Listener, 70*(12), 377–378.

Mikolov, T., Yih, W.-T., & Zweig, G. (2013). *Linguistic regularities in continuous space word representations*. Paper presented at the HLT-NAACL.

Mulgan, G. (2006). *The Process of Social Innovation. Innovations: Technology, Governance, Globalization, 1*(2), 145–162.

Narayanamurti, V., Anadon, L. D., & Sagar, A. D. (2009). Transforming energy innovation. *Issues in Science and Technology, National Academies, 26*(1), 57–64.

Narayanamurti, V., Odumosu, T., & Vinsel, L. (2013). RIP: The basic/applied research dichotomy. *Issues in Science and Technology, 29*(2).

Narayanamurti, V. (2016). *Personal communication*.

Okoli, C., & Pawlowski, S. D. (2004). The Delphi method as a research tool: An example, design considerations and applications. *Information & Management, 42*(1), 15–29.

Paletz, S. B. F., Schunn, C. D., & Kim, K. H. (2013). The interplay of conflict and analogy in multidisciplinary teams. *Cognition, 126*(1), 1–19. https://doi.org/10.1016/J.Cognition.2012.07.020.

Peierls, R. (1960). Wolfgang Ernst Pauli. 1900–1958. *Biographical Memoirs of Fellows of the Royal Society, 5,* 174–192.

Pennebaker, J. W (2014) (2011). *The secret life of pronouns. What our words say about us*. New York, N.Y.: Bloomsburty Press. Abruf am.

Pirolli, P., & Card, S. (2005). *The sensemaking process and leverage points for analyst technology as identified through cognitive task analysis*. Paper presented at the Proceedings of International Conference on Intelligence Analysis.

Polanyi, M. (1967). *The tacit dimension*.

Salton, G. (1975). *A theory of indexing* (Vol. 18). SIAM.

Sawyer, R. K. (2011). *Explaining creativity: The science of human innovation*. UK: Oxford University Press.

Senge, P. M., & Suzuki, J. (1994). *The fifth discipline: The art and practice of the learning organization*. New York, N.Y.: Currency Doubleday.

Shore, J., Bernstein, E. S., & Lazer, D. (2014). *Facts and figuring: An experimental investigation of network structure and performance in information and solution spaces*.

Silberman, S. (2015). Neurotribes: The legacy of autism and the future of neurodiversity. Penguin.

Simon, H. A. (1956). Rational choice and the structure of the environment. *Psychological Review, 63*(2), 129–138.

Simonton, D. K. (2013). Creative problem solving as sequential BVSR: Exploration (total ignorance) versus elimination (informed guess). *Thinking Skills and Creativity, 8,* 1–10.

Singer, M. (2014). Tony Hsieh: Building his company, and his city, with urbanism. *AIArchitect, 21*.

Sismondo, S. (2011). *An introduction to science and technology studies*. NJ: Wiley.

Swan, M. (2013). The quantified self: Fundamental disruption in big data science and biological discovery. *Big Data, 1*(2), 85–99.

Toulmin, S. E. (1961). *Foresight and understanding: An enquiry into the aims of science*. USA: Greenwood Press.

Tsao, J., Boyack, K., Coltrin, M., Turnley, J., & Gauster, W. (2008). Galileo's stream: A framework for understanding knowledge production. *Research Policy, 37*(2), 330–352.

Tsao, J. Y., Emmanuel, G. R., Odumoso, T., Silva, A. R., Narayanamurti, V., Feist, G. J. … Sun, R. (2013, December, 2013). *Art and science of science and technology: Proceedings of the forum and roundtable (June 5–7, 2013)*. Albuquerque, New Mexico: Sandia National Laboratories.

Uzzi, B., & Spiro, J. (2005). Collaboration and creativity: The small world problem1. *American Journal of Sociology, 111*(2), 447–504.

Waber, B. (2013). *People analytics: How Social sensing technology will transform business and what it tells us about the future of work*. NJ: FT Press.

Wilson, E. O. (2012). *The social conquest of earth*. Ney York, NY: WW Norton & Company.

Chapter 15
Knowledge, Skill, and Wisdom: Reflections on Integrating the Social Sciences and Engineering

W. Bernard Carlson

Throughout the world, individuals and nations now depend on complex technological systems to provide food, shelter, energy, and information, and these systems are largely the product of engineering expertise. In building these systems, engineers have developed a deep knowledge of the forces of nature, and they sustain and disseminate this knowledge by organizing it into specialties. Yet as our dependency on engineering increases, fewer and fewer laymen—or even engineers themselves—have a grasp on the overall nature of engineering knowledge. Concentrating on their specialties, engineers have not been called upon to articulate the basic intellectual assumptions of their discipline, to describe the social and cognitive processes by which they create and apply new knowledge, or to explain to the public the role of engineering in society. Caught between dependence and specialization, lay people find themselves increasingly frustrated and suspicious of engineers and technology; while they find more computers and machines in their lives, lay people feel that they know less and less about technology. At the same time, engineers continue to use specialization to organize their knowledge, but they are often troubled that their expertise does not translate into the social, economic, and political power needed to bring about technological progress.

15.1 Knowledge, Skill, and Wisdom

In order to address the tensions surrounding dependency and specialization, my colleagues and I in the School of Engineering and Applied Science at the University of Virginia (UVA) believe that engineers need to become reflective practitioners

W. B. Carlson (✉)
Engineering and Society Department, University of Virginia,
Charlottesville, VA, USA
e-mail: wc4p@virginia.edu

E. Subrahmanian et al. (eds.), *Engineering a Better Future*,
https://doi.org/10.1007/978-3-319-91134-2_15

who understand both the power and the limits of their professional expertise (Schon 1984). In our view, reflective practitioners in engineering are individuals capable of thinking about the nature of engineering knowledge. This does not mean that engineers need to formulate a highly complex epistemology. Instead, one of my colleagues, Gorman (1998), has found it useful to think about engineering in three simple categories: knowledge, skill, and wisdom.

In Gorman's framework, knowledge refers to the "facts"—the information—that engineers use in solving problems. Knowledge may include equations, theories, and models as well as specific devices and circuits. Yet it is not enough to possess engineering facts; one must also know how to use them. Gorman uses the term skill for knowing how to employ information. For example, in most engineering schools, students are taught how to represent complex natural phenomena in terms of mathematical equations, how to solve those equations, and to use the solutions to predict the future state of the phenomena. Equations are the knowledge, while skill is knowing how to apply the equation and evaluate the results. Of course, to become effective engineers, students need to acquire not just mathematical virtuosity but other skills such as design protocols, laboratory procedures, sketching, writing, and speaking (and I'll say more about this later when I talk about representation).

But beyond information and skill, engineering students need wisdom. By wisdom, Gorman means that students should learn when and why it is appropriate to apply their skills. Students need to know when representing a problem mathematically will yield useful information and a valid prediction. They need to be sensitive to how important aspects of natural and social phenomena cannot be measured or quantified and hence not included in a mathematical model. Students need to have the wisdom to ask what implicit assumptions may be built into a theory, model, or software package, assumptions that could lead to erroneous and even disastrous results. And they should be able to think about the larger social and cultural goals to which they are applying their expertise. As Paul Goodman once observed, "[T]echnology … aims at prudent goods for the commonweal …. As a moral philosopher, a technician [or engineer] should be able to criticize the programs given him [or her] to implement" (Martin and Schinzinger 1989, 1). Wisdom is being able to discuss which goods are prudent, what constitutes the common wealth of a society, and which engineering practices will contribute to the good society.

Wisdom may seem like an overblown and presumptuous term, but I think it is exactly the right word for what we are talking about here. Wisdom captures the mysterious quality which separates the expert from the novice, the engineer from the technician—judgment. We depend on all professionals to exercise judgment—to be able to make decisions about what to do and to acknowledge that decisions have consequences. The word wisdom elevates judgment beyond mere skill, reminding us that those entrusted with knowledge and skill must exercise their power thoughtfully, carefully, and ethically.

My purpose in introducing this three-part model of engineering knowledge is to make a simple point: we engineering educators spend far too much time cramming knowledge and skills into the heads of our students. In my opinion, much of what passes for reform engineering education today in America simply concerns increasing the amount of knowledge and skills that will be imparted to our students. We naively assume that the students will somehow acquire wisdom on the job or perhaps through divine intervention. Unfortunately, students tend to value only what we teach them. If we only teach our students knowledge and skills and neglect wisdom, then many of them will not appreciate the importance of wisdom. If we want engineering students to become reflective practitioners—capable of producing prudent goods for the commonweal—then we need to devote a portion of the undergraduate curriculum to cultivating wisdom.

15.2 An Overview of UVA's Program in Science, Technology, and Society

So, how then might we go about cultivating wisdom in engineering education? At UVA, we help engineering students to become reflective practitioners by teaching a series of required undergraduate courses which combine instruction in writing and speaking with a discussion of the nature of knowledge and the interaction of technology and society. Let me turn now to providing a brief overview of the University of Virginia and the Program in Science, Technology, and Society (STS). Following this overview, I will then suggest several of the key concepts from the humanities and social sciences which we strive to convey to our undergraduate engineering students at Virginia.

A bit of institutional context is perhaps useful here. By American standards, the University of Virginia (UVA) is a medium-sized state university with a total enrollment of 24,000. The School of Engineering and Applied Science is small by American standards, enrolling 3700 students in its undergraduate (2800) and graduate (900) programs. The engineering faculty totals 160 tenure-track faculty, and the STS program has 7 tenure-track and 7 teaching faculty. STS has long been a part of UVA Engineering and was founded in 1932 as Engineering English. At that time the university was highly decentralized, and engineering students could only take courses in the School of Engineering; they could not enroll in courses in the College of Arts and Sciences. Consequently, when the engineering faculty decided that their students should have instruction in writing, speaking, and the humanities, the logical step was to create a program inside the Engineering School. In the 1960s, the Division began offering courses in literature, history, aesthetics, and photography, and its name was changed to the Humanities Division. In the 1990s, as the Division acquired a stronger research program in STS and the history of technology, so it was renamed the Department of STS. In 2012, the STS program joined with Applied Mathematics and technology entrepreneurship to create the Department of Engineering and Society. Today, the STS program has a diverse faculty, with specialists in literature, history, anthropology, psychology, and sociology.

Over the years, the STS Program has pursued two basic missions: first, to teach engineering students to write and speak effectively; and second, to inspire students to think in rich and sophisticated ways about the links between technology and culture. The Division achieves these goals by offering four courses which are required for all undergraduate engineers. In the first semester, students take STS 1500 which covers the fundamentals of writing and speaking for professionals. The course also introduces them to engineering as a profession and a discipline. In STS 1500, we argue that writing and speaking are as important as mathematical skills for the engineers, and we suggest how one must need to use these skills flexibly and ethically.

During their second or third year, the students take a 2000- or 3000-level course on technology and society. The Program offers a range of these courses and they reflect the research interests of the faculty. We offer several courses on the history of technology, but in addition we also teach courses on utopian thinking, literature and technology, social theory of technology, creativity, and entrepreneurship.

While the topics vary, the thrust of these courses is to help the students to think about the ways in which people shape technology to reflect social and cultural goals. At the same time, we use these courses to continue to sharpen the students' communication skills, and we require the students to write several extended essays, give speeches, and analyze key texts from the humanities and social sciences.

In their fourth year, the students take a two-course sequence, STS 4500-4600, in which they write their engineering theses. The undergraduate thesis is an independent engineering project that demands creativity, long-term planning, and professional responsibility. Each student selects a topic, pursues research in the laboratory, library, or field, and then prepares a portfolio consisting of a technical report and an STS analysis of the research undertaken. Although the students often investigate highly technical topics, UVA Engineering prides itself on turning out engineers who can communicate with a variety of audiences, and this goal is achieved by having STS professors serve as the principal thesis advisers. Consequently, much of the 4500-4600 sequence is devoted to helping students design their projects and to honing their writing skills, but a portion is also devoted to discussing engineering ethics. Overall, the 4500-4600 sequence helps our students move from being passive consumers of engineering knowledge to being active producers of knowledge who can solve problems and create new designs.

Much of our energy in the STS Program is devoted to undergraduate engineering education. However, this does not mean that my colleagues and I do not have active research programs. Indeed, our undergraduate teaching has stimulated many of us to step up our research and writing efforts. Because we are deeply involved in the education of engineers, we often see how concepts and theories from STS do not fully explain the processes by which engineering knowledge is generated and transmitted. By being at the point of knowledge production (to use Bruno Latour's phrase), my colleagues and I often see issues and puzzles that stimulate our ongoing study of technology.

Moreover, the engineering students often pose questions about technology and society which cannot be answered by simply referring them to some existing book or article; all too often, the STS literature is written for a highly specialized audience of scholars in the humanities and social sciences and the literature is not accessible to professional engineers. And finally, by being an interdisciplinary group in a professional school, my colleagues and I stimulate each other to think outside the standard disciplinary boundaries and to frame research problems in new ways. As a result of all of this stimulation, my colleagues and I have a variety of ongoing research projects which have been supported by grants from the National Science Foundation, the National Endowment for the Humanities, the Social Science Research Council, and the Sloan Foundation.

15.3 Key Concepts

With this overview in place, let me turn now to the ideas we try to impart to our students in this curriculum. I want to explain these themes because, in many ways, it may be the ideas, not the curriculum, that can be transferred from one institution to another. As an aid to learning, I have condensed these themes into aphorisms, but of course, these slogans do not fully capture all the nuances of what we try to convey to the students.

(1) To represent is to know. This slogan embodies a fundamental assumption about how humans learn about the material world and gain mastery over it. As the cognitive scientist, Donald Norman (1993), argued, humans solve problems by creating different kinds of representations. By picturing a problem using different images, words, symbols, or numbers, we are able to isolate key factors, identify patterns, and come up with solutions. Frequently, people use a range of different representational techniques—taking notes, making sketches, writing equations—to acquire, and organize information about the world. In engineering education, we pride ourselves on teaching students a host of representational techniques: complex mathematics, powerful computer models, precise ways of drawing, and rigorous ways of writing and speaking. We teach these techniques so that students can analyze, predict, and eventually control the forces of nature. Yet, while engineers intuitively know that the power of their profession is grounded in their ability to represent the natural world, we educators frequently fail to convey this essential lesson to our students (Carlson 2003).

Consequently, in our STS courses, we strive not only to improve how our students represent technology through writing and speaking but we call their attention to the central role that representations play in technological problem-solving. We argue again and again with students that if you cannot describe a problem (i.e., represent it) then you will be hard pressed to solve it. As the novelist Robert Penn Warren used to say, "Clear writing is straight thinking." In making this argument, we are making a radical claim about engineering, emphasizing that communications is not ancillary to engineering but rather at the heart of engineering.

In my own first-year communications course, I developed the theme of knowing and representing by having the students build a pendulum clock and then undertaking a series of written and oral assignments related to the clock. In order to appreciate how engineers use words and images to think, I have them revise the instructions, prepare an illustrated technical description, write a patent for an improvement on the clock, and prepare a manufacturing and marketing proposal (Carlson and Peterson 1996). In doing so, I was trying not only to improve the students' communications skills but also introduce a way of thinking about engineering. In particular, I suggested that they view their undergraduate education as a process by which they will master several different modes of representation—mathematics, computer models, visual images, and words—and that the challenge is learning how best to apply these different modes to mastering natural and social phenomena. My students reported that this is often the one "big idea" they learned in their first year of engineering, and that it helped them to make sense of the overall education process.

The next two aphorisms embody what we try to teach the students about the interaction of technology and society.

<u>(2) No widget advances without social allies; hence all engineers are sociologists.</u> In this slogan, "widget" refers to any technological artifact, device or system, and I am revealing that much of what we teach reflects the actor-network approach in STS today. As Law and Callon (1992) argued, the successful introduction of a new widget depends on how well engineers are able to convince other social groups to produce and use the new widget. To use a familiar example from the history of technology, when Thomas Edison developed his incandescent lighting system in the late nineteenth century, he not only had to invent a new light bulb but he also had to convince capitalists to lend him money to finance his research, persuade the Patent Office to grant him patents for his inventions, secure permission from the government of New York City to lay cables under the streets, and convince consumers that his incandescent lighting was preferable to gas lighting (Hughes 1983; Bazerman 2002). In developing this and other examples with our engineering students, we strive to make the point that the development of the technology (i.e., the incandescent lamp) was not enough to create an electrical revolution; Edison had to be willing to negotiate and enroll a variety of groups into his socio-technical network. To succeed with these processes of negotiation and enrollment, Edison had to understand the structure of his society and the values of different groups. In this sense, Edison was a sociologist, and for the same reason, we argue that our students need to understand the structure and values of contemporary society. If UVA students wish to make sure that the widgets they design are put into widespread use, then they need to know something about the processes by which social groups appropriate technology.

As you might expect, we cover how engineers and society negotiate new technology in our 2000- and 3000-level courses. Likewise, it should be readily apparent how this theme is tied to our communications mission–engineers can only recruit various groups by effectively communicating with them. To enroll groups in their networks, engineers must be able to describe technology clearly for non-technical audiences and link the technology to the values of the target groups.

But even more than this, we use the undergraduate thesis as a hands-on exercise in building a socio-technical network. For instance, when I teach STS 4500, I introduce Law's (1987) notion of heterogeneous engineering, emphasizing that the success of a technological project depends on how well the engineer is able to hold together a series of technical and nontechnical elements. I then have the students identify the range of heterogeneous elements they need to bring together in their thesis project. I ask them to identify the equipment they will need, the people they will enroll (an engineering professor as technical adviser, support technicians, graduate students, and me), and the information and ideas (found in the engineering literature) which will justify and bolster their project.

In their project proposals, the students have to argue they can create this network and use it to accomplish their goal (i.e., getting new laboratory results or designing a new widget). In framing their thesis as a networking-building effort, I find that the students not only write better theses but they also begin to acquire the organizational skills they need to be effective leaders and managers.

(3) A society gets the technology it wishes for. This slogan is a modification of the old adage "A society gets the politics it deserves." This aphorism is an important supplement and corrective to the previous slogan. In the course of negotiating with social groups, it would be easy for students to come away with a sense that the negotiations related to technology turn on explicit, utilitarian values and issues. To return to the electric lighting example, all parties might agree with Edison that if his incandescent lighting system allowed for more illumination at cheaper prices, then everyone in society benefited—more street lights made the city safer, the middle class could enjoy lighting in their homes and businesses, workers got jobs in power plants and electrical equipment factories, while Edison and the capitalists earned a return on their investment.

But on another level, the negotiations often involve implicit cultural values—things for which a society wishes. Electric lighting "took off" in America in the late nineteenth century because it was perceived as being clean and modern, electrical utilities were seen as not being as corrupt as gas companies, and electrical systems provided a way of unifying communities, cities, and regions that might otherwise splinter along political and economic cleavages (Nye 1990). In our courses, we help students to appreciate how individuals such as Edison sometimes see and integrate these values into their socio-technical networks. At the same time, we remind students that often historical actors do not anticipate how their technological inventions will be linked to cultural values, and how these unexpected linkages often shift a technology away from its explicit utilitarian "bottom line". To explore the wishes that different societies have for technology, we often turn to literature, and our students read novels such as *Frankenstein*.

The final slogan summarizes our view on the role of ethics in engineering: (4) Engineers cannot just throw technology over the wall. Consumers may choose to use technology in unexpected ways, and hence engineers need to exercise moral imagination. The first part of this aphorism reflects the danger and conceit of specialization in engineering. It is all too easy for our engineering students to say, "I'll do my bit–design my widget, but it is someone else's problem to figure out how it will be used." The obvious problem is that this attitude leads engineers to design technological artifacts that are not only difficult to use but in fact can have serious unintended consequences (Tenner 1996).

My colleagues and I believe that the best way to overcome this problem is to help students to develop their moral imaginations. According to Werhane (1996), a professor of business ethics at UVA, moral imagination refers to the ability of professionals to imagine a variety of outcomes for their decisions. Werhane emphasizes that if one is not able to imagine different scenarios, then one cannot assess the risk or apply a framework for moral reasoning (such utilitarianism, Kantian duty ethics, Lockean Right Ethics, or Aristotelian virtue ethics). Many of us have found that the most effective way to cultivate moral imagination is to teach cases similar to those used in business schools; these cases provide students with detailed background information and then challenge them to frame an ethically appropriate action. Working with a team of graduate students, Gorman et al. (1999) developed a series of cases which examine the ethical issues surrounding the design of environmentally sensitive products. Used in our STS 4500–4600 sequence, we find that these cases help students to clarify their views about the environment and practice applying different ethical theories to real-world situations.

In teaching Gorman's cases as well as other cases from business schools, another colleague, Rosanne Welker, and I (2001) have found that we can expand the moral imagination of students by getting them to consider how technologies can have four kinds of consequences: there can be good and bad consequences and there are intentional and unintentional consequences. In everyday life, engineers earn their keep by maximizing the positive intended consequences of a technological system while minimizing the negative intended consequences. If a technological system has unintended positive consequences, then these are seen by society as a bonus, and the engineers are treated quite suitably as heroes or geniuses. However, this leaves a fourth class of consequences, the unintended negative, with which no engineer wants to deal. Welker and I, however, argue that the ultimate challenge for an engineer is to have the character and courage necessary to imagine the unintended, negative consequences of his or her design. Moreover, we discuss with the students how different groups of technologists may develop collective design practices which minimize the likelihood that a particular artifact will fall into the negative, unintended design category and hence fail catastrophically. At the moment, our discussions often turn to traditional wooden shipbuilding, in which general principles permitted a certain amount of variation and innovation but at the same time generally ensured that most boats were safe from structural failure (Chapelle 1935).

Likewise, we debate with our students how centralized, hierarchically structured systems (for computing and telecommunications) have to be carefully designed so as not to have serious negative unintended consequences and crash catastrophically. To return to the beginning, the wisdom we would like our engineering students to have is to possess a moral imagination.

15.4 Neither Preach nor Apologize

These four aphorisms represent much of what we think and teach every day in the STS program at UVA. We believe that if we can convey these "big ideas" to the engineering students, then we are succeeding in integrating the humanities and social sciences into engineering education. These ideas, we hope, will permit our students to become reflective practitioners who will be capable of thinking about and directing technology toward the goals of the next generation. If we do our job well, then we are not just "humanizing" the engineers; indeed, we are creating a new breed of engineer.

But while my colleagues and I have lofty goals for our approach to teaching engineering students, we are nonetheless sensitive to the importance of not being arrogant. Over the years we have learned to try not to preach to the engineering students, taking the view that somehow we humanists have a broader and morally superior view of life. As tempting as it is to see the classroom as the opportunity to win converts for the First Church of Actor-Network Theory, we instead see our task as drawing on ideas from the STS literature and the history of technology in order to frame questions which help the students to think about their future role in society.

At the same time, my colleagues and I struggle not to be apologists for the status quo in engineering. All too frequently, people think that we should just teach the history of technology to undergraduates in order to demonstrate how great white men did great things with great machines and that this sort of history proves that we live in the best of all possible technological worlds. No, our task as humanists and social scientists who teach in an engineering school is not to confirm that the status quo. Rather our tasks are to prepare students who can imagine a better world and to provide them with the skills and wisdom that will permit them to create it.

References

Bazerman, C. (2002). *The languages of Edison's light: rhetorical agency in the material production of technology*. Cambridge: MIT Press.

Carlson, W. B. (2003). Toward a Philosophy of Engineering: The Fundamental Role of Representation. In *Proceedings of the american society for engineering education*, CD-ROM.

Carlson, W. B., Peterson, K. (1996). Making clocks: a first-year course integrating professional communications with an introduction to engineering. In *ASEE Proceedings*, CD-ROM.

Carlson, W. B., Welker, R. (2001). The whammy line as tool for fostering moral imagination. In *Proceedings of the american society for engineering education*, CD-ROM.

Chapelle, H. I. (1935). *The history of American sailing ships*. New York: W. W. Norton.

Gorman, M. E. 1998. Transforming nature: ethics, invention, and discovery. Boston: Kluwer Academic Press.

Gorman, M. E., Mehalik, M., Werhane, P. (1999). *Ethical and environmental challenges to engineering*. Prentice-Hall.

Hughes, T. P. (1983). *Networks of power: electrification in western society, 1880–1930*. Baltimore: Johns Hopkins University Press.

Law, J. (1987). Technology and heterogeneous engineering: the case of portuguese expansion. In W. E. Bijker, T. P. Hughes, T. J. Pinch (Eds.), *The social construction of technological systems* (pp. 111–134). Cambridge: MIT Press.

Law, J., Callon, M. (1992). The life and death of an aircraft: a network analysis of technical change. In W. E. Bijker and J. Law (Eds.), *Shaping technology/building society: studies in sociotechnical change* (pp. 21–52). Cambridge: MIT Press.

Martin, M. W., Schinzinger, R. (1989). *Ethics in engineering* (2nd edn). New York: McGraw-Hill.

Norman, D. (1993). *Things that make us smart: defending human attributes in the age of the machine*. Reading, Mass.: Addison-Wesley.

Nye, D. E. (1990). *Electrifying America: social meanings of a new technology, 1880–1940*. Cambridge: MIT Press.

Schon, D. A. (1984). *The reflective practitioner: how professionals think in action*. New York: Basic Books.

Tenner, E. (1996). *Why things bite back : technology and the revenge of unintended consequences*. New York: Knopf.

Werhane, P. H. (1996). *A Note on Moral Imagination*. Case No. UVA-E-0114, Darden Graduate School of Business Adminstration, University of Virginia.

Chapter 16
Dealing with the Future: General Considerations and the Case of "Mobility"

Georges Amar

To tell is not to foretell.

Prospective is not prediction.

Big data and computers, so-called artificial intelligence, have made our predictive capacities higher than ever. Health expectancy, chance of marriage and divorce or even fortune: Everything is computable. Is it really interesting?[1]

Prediction means to simulate the development of a system out of currently available knowledge, under the hypothesis that no new element or conceptual break will happen between the prediction and its realization. To predict the future is clearly possible, and it is the best way to overlook the *new*, to deprive yourself of the «real» future, i.e., an *open* one, and not a mere repetition of the past.

However, this openness is not granted. It is a hard work to open a future "framed" by our distrust of uncertainties and insecurity, by so many inclinations stemmed from cultural as well as material heritage. This is how I see and practice prospective: not to determine the future but on the contrary to «under-determine» it, as philosophers say, or as engineers or physicians would put it, to enhance its degrees of freedom. The problem of the future is not a lack of knowledge. It is rather that the future is determined by powerful mechanisms of all kinds, including mental ones, in which we are caught. For instance, it is «written» that the global warming will pass the dangerous threshold of 2 or 3 degrees.

---

[1]It is certainly useful, for weather forecast, insurance, risk prevention, and preparation of various events, at least the ones you are able to envisage, hence to predict. For the rest, it would rather heighten the risk of missing it all.

G. Amar (✉)
Mines ParisTech-PSL Research University, Paris, France
e-mail: amar.georges26@gmail.com

G. Amar
RATP, Paris, France

E. Subrahmanian et al. (eds.), *Engineering a Better Future*,
https://doi.org/10.1007/978-3-319-91134-2_16

The aim of prospective is not only to make that sort of prediction, or to contest it, but to offer new perspectives. To reopen the game.

How does it work concretely? The prospective exercise has two faces. First, a critical study of current paradigms in a specific domain. A prospective criticism consists in identifying the (or a) dominant paradigm in this domain and to evaluate its degree of obsolescence: Are its concepts and denominations system still adequate? To which extent does its unfitness hinder our capacity to appreciate the emerging realities and to welcome the nascent?

The other face of the prospective exercise is the detection and the formulation of emerging paradigms in the given domain. The prospective formulation is a creative act as well as an awareness of something that is already there, implicitly or confusingly. Its expression is an effective way to help trigger the emergence of a paradigm. To name a new paradigm soundly, in a clear, understandable and yet uncommon (even enigmatic) fashion is a condition of its fertility and thus determines the strength of the *prospective proposition*.

This twofold exercise, both critical and «pro-positive», demanding expertise and creative conceptualization, characterizes the *conceptive prospective* thus named to oppose the predictive one. It can be seen as a Design activity that does not target new products but new conceptual fields. It calls for various knowledge, but in order to nurture a work on concepts and languages. This work, which nears literature, art or philosophy, can regenerate the unknown, and «re-open» the future, as we have said earlier. The role of the prospective is neither to distress nor to reassure, it is to stimulate and sustain the creativity proper to innovators, strategists, all those who love the future!

Time to highlight this theoretical approach through a concrete domain.

It will be «Mobility», for the following reasons.

The first one is that I know it well enough. I spent a long career in one of the major urban transport companies (RATP, multimodal public transport system of greater Paris, France). I had various activities, in production as well as in the management of major projects. I worked as a researcher, and was in charge of the prospective and innovation department. I had the opportunity, over a long period (35 years), and from many angles, to witness the great changes of the urban transport domain, to try to understand them and to anticipate their evolutions.

Second reason: the study of mobility can be fruitful for other domains as it offers a good illustration of a conceptive prospective approach.

It is not exaggerated to say that the standard transport paradigm has entered a phase of obsolescence. «Transport» is the (multiple) answer to *this* question: «How to go the fastest possible (and in the best conditions of security, cost, quality, comfort) from point A to point B?» The emerging paradigm that has started to overcome it (or to complete it) is that of the «Mobile Life». It is a conceptual rupture that deeply affects our habits and ways of living, economic actors and business models, services, tools, and infrastructures. The iconic tool of this passage to the mobile life is no doubt the smartphone. It is noticeable that this device is now what we call a *mobile*. Therefore, automobile, airplane, and train have become the symbols of the «ancient mobility» (*transport*)—which does not mean that they are going to disappear (after all the horses are still here!). In fact, these three classic modes, beacons of the twentieth century, will do better than survive; they will undergo a profound mutation. The car will see a brighter future! However the car of the future (particularly in the horizon of its autonomy) will not be «the car» in the sense we are giving it today, inasmuch as a smartphone is not at all a «telephone» (a tool whose definition is to allow speaking from a distance), although one can use it to telephone on occasion. The smartphone is a tool of the mobile life. Of the mobile Internet; of mobile trade, education, health, leisure, and even of mobile love. The car of the future will no longer fit in the category of the *vehicle* (a means to go from A to B) but to the categories of, somehow odd to our ears, «mobile place» and «mobile body».

Although the smartphone stands out as the superstar of the mobile era, it has, especially in the prevailing urban context, two unexpected but significant companions: the bike and the shoes!

The bike, pleasant, convivial, excellent for the health and for the environment, qualified as a mode both *soft* and *active*, has become within a decade, a real urban «must». No modern city can do without biking tracks, services of Velib type (short rental self-service bicycle), stations, electrical bikes. But beyond the «velocipede», a compound word made of the latin *velox* (rapid) and the *biped*, it is the walking human body (the biped) that is making a major comeback in the new and next mobility. The body, equipped, «augmented», not only with shoes and high-tech outfits, will be the subject of stunning innovations to come. The «oldest transport mode ever» is likely to be the most promising one for the future.

Walking is back, under different forms! The new technological objects are expanding: gyropodes, urban stilts, electrical monocycles, and various peculiar urban gliding engines. Walking is also giving us a low-tech example of one of the most fertile innovation process today: hybridization. One of the cutest inventions of this type is the «Pedibus» (or Walking Bus) that can take the kids to school, exactly in the same manner as a bus (fixed route, stops, schedule, driver) except that there is no "bus" (vehicle)! The Pedibus is a software without hardware, a fantastic «crossing» between walking and bus.

This kind of conceptual hybridization has expanded in the world of transport. Hence the classic conceptions of «public» versus «individual» transport gradually subsides, making room for a new paradigm, unusual yet already in practice: the «PIT». I have proposed this contradictory expression («Public Individual Transport») for some times now, in order to focus on the emergence of an innovation field the richness of which is indubitable today. Its exceptional development (BlaBlacar, Zipcar, Velib, Uber, Drivy, and many others) has dismantled the solid partitions between the public/common/individual/private which our public policies (namely about "modal shift") have stranded on for decades.

Several other hybridizations have surfaced at the beginning of the twenty-first century, for example, that of Bus and Metro (the famous Brazilian born BRT—*Bus Rapid Transit*) or that of Tram and Train (in Germany). The most outstanding however are and will be the crossings of the «physics» and the «digital». The world of Video games are already giving disconcerting examples, with for instance the Nintendo Wii consoles (how to play "real" Tennis in your living room?), and various ones based on the Kinect of Microsoft and other movement capture tools. The most striking example of hybridization in transport is the awkwardly named 3D Printer: a form of *material* mobility with no physical motion! These are yet a few prefigurations of the future mobility hybrids.

Finally, the third conceptual prospective interest in the mobility field is that its future is still more widely open than what we have just acknowledged. One should not believe that the future of mobility, as our prospective approach has drawn, means a generalized restless agitation, a perpetual jactitation, exhausting, less and less efficient. We only begin to understand that the «real» mobility is not only a matter of miles and miles per hour, but also of connectedness and «reliance» (French word to say creation of links, discoveries, encounters, opportunities and serendipity). It is a matter of physical, mental and social health, and also of pleasure and leisure. We begin to understand that the «true» mobility is a mode of expression (such as your personal walk style), a very rich multisensorial experience embracing esthetic and poetic dimensions.

Finally, we begin to see that the notions and experiences of *mobile* and *immobile* are not opposed, nor even entirely distinctive. After all, when we are «sitting quietly doing nothing» (typical zen meditation quote) we swing along with our good planet in its movement on itself and in the universe, at an unbelievable speed.

www.ingramcontent.com/pod-product-compliance
Lightning Source LLC
LaVergne TN
LVHW082227200926
848484LV00008B/2313